Lectures on White Noise Functionals

LECTURES ON WHITE NOISE FUNCTIONALS

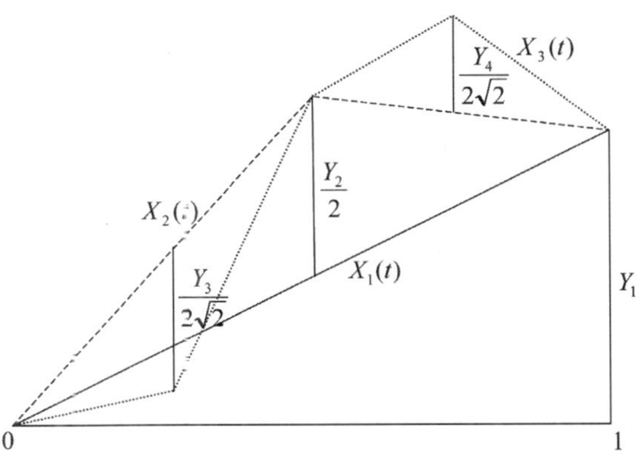

T Hida • Si Si

World Scientific

NEW JERSEY • LONDON • SINGAPORE • BEIJING • SHANGHAI • HONG KONG • TAIPEI • CHENNAI

Published by

World Scientific Publishing Co. Pte. Ltd.
5 Toh Tuck Link, Singapore 596224
USA office: 27 Warren Street, Suite 401-402, Hackensack, NJ 07601
UK office: 57 Shelton Street, Covent Garden, London WC2H 9HE

Library of Congress Cataloging-in-Publication Data
Hida, Takeyuki, 1927–
Lectures on white noise functionals / by T. Hida & Si Si.
p. cm.
Includes bibliographical references.
ISBN-13: 978-981-256-052-0 (hardcover : alk. paper)
ISBN-10: 981-256-052-1 (hardcover : alk. paper)
1. White noise theory. 2. Gaussian processes. I. Si Si. II. Title.
QA274.29.H53 2004
519.2'2--dc22

2008003893

British Library Cataloguing-in-Publication Data
A catalogue record for this book is available from the British Library.

Printed in Singapore by World Scientific Printers

To U Khin Aung and to the Memory of Tsuta Ohta

Preface

These notes are based on a series of lectures delivered by one of the authors for the infinite dimensional analysis seminar at Meijo University in the last five years. They are presented here as lectures on white noise functionals. The aim of the lectures are twofold; one is to explain the basic idea of the white noise analysis to those who are interested in this area and the second is to propose some new directions of our theory to the experts who are working on stochastic analysis. For these aims, a perspective of the white noise theory here is presented in this book.

Thus, the reader can see that effective use of our favorite generalized white noise functionals does meet the needs of the time. Indeed, it is now the time to use *generalized white noise functionals* which has been introduced as a natural development of the stochastic analysis. They serve profound roles also in quantum as well as statistical mechanics. Another basic concept of our analysis is the *infinite dimensional rotation group.* With the use of this group we can carry out a harmonic infinite dimensional analysis.

There are several aspects of white noise analysis. Among them,

1) Analysis of stochastic processes (historical reference: J. Bernoulli, Ars Conjectandi (1713) CAP II, where one can see the term *stochastice.* We are stimulated by such a classical idea.

2) The method of the theory of functional analysis developed in the second half of 20th century. This has a close connection with what we are going to discuss, although in a naive level.

3) Harmonic analysis arising from

i) the infinite dimensional *rotation group*, and
ii) the infinite symmetric group.

There are interesting relationships between the two groups.

4) Following the standard roadmap of analysis, first comes determination of basic variables to define functions (in fact, functionals) to be dealt with. Then follow the analysis and applications.

One of the *characteristics* of what we are studying is that the space of variables is *infinite dimensional*, so that we discuss an infinite dimensional analysis. As for the infinite dimensionality with which we are concerned, one should not be simple minded. The reason why we have to be so careful to deal with this concept will be explained in many places in these notes by taking actual objects, however we now explain in an intuitive manner.

In order to analyze random complex systems, which are the object to be investigated, the basic variables are preferable to be independent, stationary in time and atomic. Good examples are (Gaussian) white noise $\dot{B}(t)$ and Poisson noise $\dot{P}(t)$. They are the time derivatives of a Brownian motion $B(t)$ and a Poisson process $P(t)$, respectively. Both are systems of *idealized elemental random variables.* Independence at every t comes from the fact that each of those two processes has independent increments.

To fix the idea, let us take white noise $\dot{B}(t)$. The time parameter t runs through a continuum, say R^1. Since white noise $\{\dot{B}(t), t \in R\}$ is a system of continuously many independent random variables, although each member is an infinitesimal random variables, the probability distribution might be thought of as a direct product of continuously many Gaussian distributions. Separability might not be expected. Obviously this is not a correct understanding as we shall see in Chapter 2. In fact, sample functions of $\dot{B}(t), t \in R$, are generalized functions, so that the probability distribution of white noise is given on the space of generalized functions. The measure space, thus obtained, is certainly separable. These two facts seem to be a contradiction, but in reality, it is not so.

An interpretation of these facts will be given in Chapter 2, where an identity, as it were, of $\dot{B}(t)$ is given as a generalized white noise functional (indeed, it is a linear generalized functional).

Note that white noise $\dot{B}(t)$ has been understood as a generalized stochastic process, from a view point of classical stochastic analysis, where smeared variables such that

$$\dot{B}(\xi) = -\int B(t)\xi'(t)dt$$

have meaning.

We now propose to take $\dot{B}(t)$ without smearing. Indeed, white noise analysis does not want to have white noise smeared, since the time variable t should not disappear. Having fixed the system of variables $\{\dot{B}(t), t \in R\}$, we come to nonlinear functionals, starting from polynomials in $\dot{B}(t)$'s. Unfortunately, bare polynomials (with degree ≥ 2) cannot be in any acceptable class of white noise functionals. We have therefore found a most significant trick, which is called *renormalization*, to make polynomials to be in a class of acceptable functionals. The class is nothing but the space of generalized white noise functionals that has been emphasized at the beginning of this preface. This class will be discussed in details in Chapter 2. Indeed, with the use of generalized white noise functionals, we have done more than classical stochastic analysis.

As the next step, we are led to introduce a partial differential operator in a generalized sense such that

$$\partial_t = \frac{\partial}{\partial \dot{B}(t)},$$

(see Kubo-Takenaka[89]).

Other notions and operators like Laplacians that are necessary for our analysis can naturally be introduced successively.

We should note that Poisson noise and its functionals can be discussed in the same idea, but important fact is that dissimilarity between two noises is quite interesting and important. Furthermore, we have discovered some *duality* that provide connections between them.

Thus, a roadmap is ready and a new stochastic analysis proceeds. Needless to say, we can establish intimate relations with other fields, not only within mathematics but wider area of sciences.

These notes have been prepared so as to be self-contained. As a result, some elementary description is included. Beginners would find it easy to follow the edifice of white noise theory.

The present work, though far from claiming completeness, aims at giving an outline of white noise analysis responding to growing interest in this approach. It is our hope that the present notes would contribute to the future development of stochastic analysis and of infinite dimensional analysis.

Acknowledgements The authors are grateful to Ms. E. H. Chionh at World Scientific Publishing Co. who has helped us while we were preparing the manuscript.

February 2008, at Nagoya.

Contents

Chapter 1

Introduction

1.1 Preliminaries

We shall start with a brief preliminary to explain why we have come to the study of white noise functionals.

The authors are sure that everybody who is interested in probability theory, more generally in science, is quite familiar with Brownian motion and knows how it is involved in the theory of stochastic processes and stochastic analysis. It is therefore, natural to provide a survey of Brownian motion first of all and to explain a theory of analysis of its functionals (Brownian functionals).

Indeed, the analysis of them has been extensively developed. While we are studying such an analysis, we have the idea of taking *a white noise*, the time derivative $\dot{B}(t)$ of a Brownian motion $B(t)$, to be the variables of functionals in question, instead of a Brownian motion itself. The main reason is that the $\{\dot{B}(t), t \in R\}$ forms a system of i.i.d. (independent identically distributed) random variables. This is an essential part for the analysis of functionals of white noise. In particular, if we meet nonlinear functionals of a Brownian motion, its expression in terms of a white noise shows a big advantage to analyze them. The i.i.d. property of the variables makes the analysis simpler and efficient.

We would like to emphasize the significant characteristics of white noise analysis. The characteristics are also considered as advantages. They are now in order.

(1) The space of generalized white noise functionals is one of the main subjects of the study. They are naturally defined since $\dot{B}(t)$'s are taken

to be the variables of functionals. The space of generalized white noise functionals is very much *bigger than the classical L^2-space* of functionals of Brownian motion. As a result, we can carry on the calculus on wider class of random complex systems, including causal calculus where time development is explicitly involved.

(2) Infinite dimensional rotation group is introduced. It describes invariance of the probability distribution of white noise. We are therefore led to a harmonic analysis arising from the rotation group. Complexfication of the group has close connection with quantum dynamics and other fields of application.

(3) Innovation approach is our favorable method of analysis. White noise $\dot{B}(t)$ is typical and Poisson noise can be discussed in a similar manner since both are systems of idealized elemental random variables. Interesting remark is that important dissimilarities are found and we can even discuss duality between two noises.

More details will be discussed in Section 1.5.

1.2 Our idea of establishing white noise analysis

We shall deal with random complex systems expressed as functionals of white noise. What we shall discuss are of Brownian motion or mostly concerned with white noise and their functionals. Those of Poisson noise will also be discussed. Some of the results of the latter case are obtained in a similar manner to the Gaussian case, but dissimilarity between them is sufficient.

Our interest is focused on those systems of functionals which are evolutional, namely those systems change and develop as time or space-time parameter goes by.

As is noted in Carleton Mathematical Notes[41], we want to establish a basic theory so that it enables us to discuss random systems in a mathematically systematic manner and comprehensively. There have so far been many directions on stochastic analysis, hence it would be fine if those approaches can be set up in a unified manner.

Another motivation is that a stochastic differential equation can be dealt with as if it were the case of ordinary differential equations, although the new analysis to be proposed would be beyond the classical functional analysis. The new stochastic analysis is based on white noise, either Gaussian or Poisson noises or even compound cases, depending on the choice of variables of functionals in question.

However, it is not natural to assume that the given system is driven by a white noise that is given in advance. Generally, starting with a given random system, we have to construct a white noise, in terms of which the given system is expressed as its functional. The white noise thus obtained is nothing but the *innovation* of the system. The notion of an innovation for an evolutional random system will be prescribed later, but at present, we roughly understand that it is a system of independent random variables, may be idealized variables, which contain, up to each time, just the same information as what the given random system gains.

1. Our idea starts, therefore, with the step of **"Reductionism"** for random complex systems. Practically, we are to meet a huge building blocks of a random system to be investigated, the constituent elements of which should be elemental, and in fact atomic. They represent basic units of randomness, something like an atom in physics, where the collection of atoms constitute a matter.

We are mainly concerned with evolutional random complex systems, so that actual implementation is to construct the *innovation* which is an elemental random system by extracting necessary and sufficient information from the given random complex system. This is the step of reduction and in fact, the first and the most important step of our mathematical approach to the study of the given random system.

The standard innovation can be expressed as the time derivative of a Lévy process. This choice is reasonable, since a Lévy process has independent increments and satisfies some continuity in time which may naturally be assumed.

2. Then follows the second step **"Synthesis"**. There the given random system should be expressed as a functional (which is non-random and, in general, nonlinear) of the innovation that has just been obtained in step **1.** the reduction. Thus, we have an analytic representation of the random complex phenomenon in question by choosing suitable functionals of the innovation that have been established.

Since those functionals are non–random, we can appeal to the known theory of functional analysis, although the variables are random.

3. Finally, we are ready to study the **"Analysis"** of those functionals, in fact, nonlinear functionals of the innovation. It may be proceeded by having been suggested by the ordinary functional analysis, however it is noted that the variable is innovation, namely elemental random variables that are mutually independent. The differential and integral calculus should be newly established. This can be done by overcoming difficulties that arise there.

It is specifically noted that we can find an infinite dimensional rotation group and we can carry on an infinite dimensional harmonic analysis arising from the group (note that we do not specify the group).

Further various applications can be discussed, and one can even see beautiful interplay between our theory and the studies of actual problems in various fields of science.

After those steps, there naturally follow interesting applications in various fields of science, some of which are going to be presented in this volume.

If we follow these three steps, it may be said that we are influenced by the historically famous literature by J. Bernoulli's article "Ars Conjectandi" appeared in 1713. In fact, such a way of thinking is closely related to the definition of a stochastic process, in reality the term *stochastice* has first appeared in that literature. Also, we would like to recommend the reader the most famous monograph by P. Lévy[102], in particular, Chapter II for definition of a stochastic process. Incidentally, the reductionism in the present terminology did start, in reality, much earlier in the book, Chapter VI, although he did not use the term innovation.

It would be meaningful to recall quickly a brief history of defining a stochastic process. However we are afraid that, when such a history is mentioned, one is usually suggested to remind the traditional method, where basic random quantity is given in advance. We should first study from our viewpoint; namely first we investigate basic elemental random systems, then we see suitable combination of those elemental systems such that they can be an innovation. Thus, follow the steps **1, 2, 3**, discussed above.

At present, we are sure that the *innovation approach* is one of the most efficient and legitimate directions to the study of stochastic processes and random fields, or more generally to evolutional random complex systems. Indeed, having had many attempts to investigate various random functions, we have come to recognize the significance of the classical idea of defining a stochastic process due to J. Bernoulli, P. Lévy, A.N. Kolmogorov, and others. We shall be back to this topic in Section 4.

We can tell some more concrete story from somewhat different viewpoint. An additive process appears, in an intuitive level, from the theory of innovation of a stochastic process. For an additive process (Lévy process), a most ground breaking theory is the so-called *Lévy's decomposition theorem*[102] (it is also called the Lévy–Itô decomposition theorem). Under some additional and in fact, mild assumptions, we are given a Lévy process, the sample functions of which are ruled functions almost surely, to squeeze the innovation out of the given stochastic process, for which an explicit decomposition formula has been given. Another interesting course of the decomposition comes from the determination of the infinitely divisible probability distribution established by A. Ja. Khinchin and P. Lévy, but we do not go further in this direction.

The time derivative of an additive process is a general *white noise*. We are, therefore, led to the Lévy decomposition of a general white noise, as a result. It claims that a general *stationary* white noise can be decomposed (up to constant) into the following two parts:

1) (Gaussian) white noise,

2) compound Poisson noise.

A compound Poisson noise is a sum (may be a continuous sum) of independent Poisson noises with different scales of jump. Thus, we may choose a Gaussian noise and single Poisson noise with unit scale to be the representatives of the set of general white noises which are *elemental*.

Some brief interpretations to elemental noises will be given in the following chapter.

One of the significant traditional problems to be discussed related to our framework on innovation is the *prediction theory*, in particular the nonlinear prediction theory. The theory has stimulated the innovation theory. Heuristic approach can be seen in the work by N. Wiener and P. Masani. We refer to the monograph[114], where white noise is playing a dominant role.

Another direction is the theory of stochastic differential equations (SDE); more generally theory of stochastic partial differential equations (SPDE) and stochastic variational equation (SVE) involving white noise. We can understand that these theories can be discussed from our viewpoint of white noise analysis.

Generalizations of these directions, some concrete problems on the theory of stochastic processes and various interplay with other fields of science have led us to a general theory of white noise analysis.

Poisson noise, which is another elemental noise, can be dealt with in the similar manner as the Gaussian case, however a Poisson noise has its proper characteristics, so that we can discuss the analysis of Poisson noise functionals with much emphasis on the difference from Gaussian case. Discovering intrinsic properties of Poisson noise is an interesting and in fact significant problem.

1.3 A brief synopsis of the book

This section gives a brief synopsis of the actual content of this monograph.

Originally we planned to write this monograph so as to be self-contained, and much has been done. Unfortunately this idea is not quite successful, for one thing the white noise theory is rapidly expanding having connections with many other fields in science.

The notion of generalized white noise functionals (white noise distributions) is one of the main topics of this monograph, Chapter 3 and part of Chapter 4 will be devoted to the study of those functionals.

The main part of Chapter 2 involves the notion of generalized white noise functionals. We take systems of idealized elemental random variables, in particular (Gaussian) white noise and Poisson noise. To fix the idea, we take white noise, a representation of which is $\dot{B}(t)$. We understand that variables are given, so natural class of functions is to be defined. Our idea is to start with polynomials in the given variables, then to introduce a general class of functions. Further we shall go to the differential and integral calculus.

These steps could be accepted by everybody, and various phenomena are expressed as functions of white noise, and they are analyzed so that the given phenomena can be identified and clarified. Since the variables are independent, we may expect the stochastic calculus can be done nicely. This is true, however, we have to pay a price. First we have to give a correct meaning to $\dot{B}(t)$'s as well as to their functions. We note that it is better to call them functionals, since they depend on $\dot{B}(t), t \in R$, which is a function of t. In addition, we require to introduce a good functional representation, because original functionals are, in general, functions of generalized functions, the sample functions of $\dot{B}(t)$. This can be done; indeed, the so-called $\mathcal{T}$- or $\mathcal{S}$-transform serves this need, and does even more.

Needless to say, Gaussian systems are sitting in the center of systems of random variables because of their very significant properties, one of which is linearity. Concerning the linearity, Poisson noise comes right after. Linear combinations of white noise and Poisson noise lead us to discuss probabilistic properties by applying linear operations acting on them.

Advantages of rotation group $O(E)$

We have the infinite dimensional rotation group $O(E)$, where E is a nuclear space. This group can characterize the (Gaussian) white noise measure. It is natural that there arises an infinite dimensional harmonic analysis from the group. It is not locally compact under the usual compact-open topology, but we may still try analogous approach. While, essentially infinite dimensional analysis can be seen with the help of the infinite dimensional rotation group. These will be discussed in Chapter 5.

Like the finite dimensional case, complexfication of the rotation group enables us to have more fruitful results. Those may be said to have come from the viewpoint of harmonic analysis. In particular we can find interesting relationship between complex white noise and quantum dynamics.

Poisson noise is interesting and significant as much as Gaussian white noise. Despite the appearances of sample functions, a Poisson process or Poisson noise has significant characteristics, although they often appear in an intrinsic manner. Details will de discussed in Chapter 7. We compare Brownian motion, which is Gaussian, and Poisson process in Chapter 2 briefly before discussing concrete results in Chapter 7 and after.

Innovation theory will support our approach that starts with reduction of random complex systems (Chapter 8). Applications to variational calculus will be seen in Chapter 9, although to some extent we have discussed in another monograph[71] by us.

We would like to emphasize intimate connections with quantum dynamics. Chapter 10 is devoted to explain some notable roads to this direction.

The Appendix involves some necessary formulas, basic notions which are not in the main stream of this book, and discrete parameter white noise, which serves a note for an invitation to white noise. Certainly it is simpler than the main stream of our story of white noise depending on a continuous parameter. For instance, measurability of functionals depending on continuously many variables, structure of a nuclear space of functions, continuously many independent random variables and so forth. It should be noted that infinite dimensional rotation group and unitary group enjoy more profound and rich structure in the continuous parameter case. In any case, discrete parameter white noise is much easier to deal with compared to the continuous parameter case.

1.4 Some general background

One may ask oneself what does a random complex system mean, the answer may be, in particular, a stochastic process $X(t)$ parametrized by the time parameter t.

We are recommended to follow the method of defining a stochastic process introduced by J. Bernoulli first and later by A.N. Kolmogorov, as will be discussed in Chapter 2. We essentially follow this idea, but in our case, characteristic functional is often used in order to define and identify a stochastic process as well as a generalized stochastic process. A characteristic functional can be defined even for generalized stochastic processes, the sample functions of which are generalized functions almost surely. A generalization to the case where t is multi-dimensional, namely to the case of a random field, is almost straightforward. To fix the idea, the parameter space T, where t runs, is taken to be R^d, $d \geq 1$, or its finite or infinite interval.

a) A well-known traditional method of defining a stochastic process starts with a consistent family of finite dimensional probability distributions. Then, the Kolmogorov extension theorem guarantees the existence of a probability measure μ on R^T, or more precisely, existence and uniqueness of a probability measure μ on a measurable space $(R^T, \mathcal{B})$, where $\mathcal{B}$ is the sigma-field generated by cylinder subsets of R^T. The measure space $(R^T, \mathcal{B}, \mu)$ is a stochastic process. The μ-almost all x in R^T is a sample function of the stochastic process. We often use the traditional notation such as $X(t), t \in T$, defining $X(t) = X(t,x) = x(t)$.

b) In fact, another more familiar and traditional method is to give a function $X(t,\omega), t \in T, \omega \in \Omega$, where $(\Omega, \mathbf{B}, P)$, with the sigma-field $\mathbf{B}$ of subsets of Ω, is a probability measure space. The function $X(t,\omega)$ satisfies the condition that $X(t,\omega)$ is $\mathbf{B}$-measurable for every t.

The condition implies that any random vector $(X(t_1), X(t_2), \ldots, X(t_n))$ has a probability distribution and the requirements in a) are satisfied. Thus we are given a stochastic process which is in agreement with $X(t,x)$ in a). Almost all x are sample functions of the process $X(t)$,

c) Suppose that $X(t,\omega)$ is given as in b) and is a measurable function of t. It may be a generalized function of t for each ω. In any case, taking a suitable space E of test functions, a continuous bilinear form, denoted by $\langle X(\cdot,\omega), \xi\rangle$ with $\xi \in E$ is well defined.

Hence, the characteristic functional $C_X(\xi)$ of $X(t)$ is defined:

$$C_X(\xi) = E\left(\exp\left[i\langle X(\cdot,\omega), \xi\rangle\right]\right). \tag{1.4.1}$$

It is easy to prove that $C_X(\xi)$ satisfies the following properties.

1) $C_X(\xi)$ is continuous in ξ,
2) $C_X(0) = 1$,
3) $C_X(\xi)$ is positive definite.

Conversely, suppose a functional $C(\xi)$, satisfying the conditions 1)– 3), is given, then we can form a probability measure μ on E^*. Then, as in a), a stochastic process exists. Its characteristic functional is equal to the given $C(\xi)$. This is the characteristic functional method of defining a stochastic process.

Existence and uniqueness of a measure will be discussed in Chapter 2.

Indeed, the above result is an infinite dimensional generalization of the

S. Bochner theorem that guarantees the relationship between probability distribution on finite dimensional Euclidean space and characteristic function satisfying the conditions 1), 2) and 3) for finite dimensional case.

We understand a stochastic process, the same for a random field, from the standpoint c), however a) and b) are also referred occasionally depending on the situation.

1.4.1 *Characteristics of white noise analysis*

Generalized white noise functionals are introduced in Chapter 2. The introduction of classes of generalized white noise functionals and their effective use are one of the main advantages of the white noise analysis. It is noted that a realization of a white noise is the time derivative, denoted by $\dot{B}(t)$ of a Brownian motion $B(t)$. The collection $\{\dot{B}(t), t \in R\}$ has been understood to be a generalized stochastic process with independent values at every t in the sense of Gel'fand. Formally, it has been viewed as a system of idealized elemental random variables. We wish to form nonlinear functionals of $\dot{B}(t), t \in R$, however each single $\dot{B}(t)$ has not been rigorously defined, although only smeared variable $\dot{B}(\xi)$ was defined. We often hear that $\dot{B}(t)$ is understood as a Gaussian process with mean 0 and covariance $\delta(t-s)$; this is no more than a formal understanding.

Key point 1

Under such a circumstance, we propose to give the *identity* to each $\dot{B}(t)$ for every t. This is rigorously done, and we can proceed to define function(al)s of $\dot{B}(t)$'s. Having been given the system of variables, it is reasonable to come to the definitions of polynomials in those variables. The trick to do this is not so easy as is imagined. The so-called *renormalization* technique is used and we naturally introduce generalized white noise functionals. Starting with the usual Hilbert space of functionals of Brownian motion with finite variance, two typical classes of generalized white noise functionals are introduced; one uses an integral representation of those functionals, where Hilbert space structure remains and the other is an infinite dimensional analogue of the space of the Schwartz distributions. Both spaces involve important examples in white noise analysis, however they are slightly different and share the roles.

Chapter 4 discusses Gaussian systems which are classes of Gaussian random variables satisfying a condition under which those variables exist consistently. If permitted, we may say that a Gaussian system has linear structure, which determines the system. Roughly speaking, a Gaussian process can be represented by a system of additive Gaussian processes, in fact those processes are modified Brownian motions. Linear analysis and Brownian motions can completely determine the given Gaussian process. The Brownian motion may be replaced by white noise.

We note that linearity and white noise are involved. Keeping the linearity, we may try to replace white noise with Poisson noise (or by compound Poisson noise). For one thing, each of these two noises forms i.e.r.v. and they are atomic system, i.e. cannot be decomposed into independent nontrivial systems. So far as linear operators are concerned, we can deal with linear processes involving linear functionals of compound Poisson process in the similar manner to the Gaussian case. These will be seen in Chapter 4.

Key point 2

We then come to the second advantage of white noise analysis; namely the use of the *infinite dimensional rotation group*. As is easily seen from the expression of its probability density function, the n dimensional standard Gaussian distribution is invariant under the rotation group $SO(n)$. We expect similar situation in the infinite dimensional case, i.e. for the case of white noise measure μ. But this is not the case. In fact, we take the group $O(E)$ of rotations of a nuclear space E, due to H. Yoshizawa [171]. It contains not only the projective limit $G(\infty)$ of $SO(n)$, but involves other rotations which have probabilistic meanings. Indeed, the subgroup $G(\infty)$ occupies, intuitively speaking, only narrow part of $O(E)$.

Starting from the introduction of $O(E)$, we can carry out an infinite dimensional harmonic analysis arising from the infinite dimensional rotation group. It is noted that a subgroup isomorphic to the conformal group plays interesting roles in probability theory.

Complexfication of the rotation group, that is the *infinite dimensional unitary group*, is denoted by $U(E)$. There appear more interesting subgroups, where significant roles are played by the ordinary Fourier transform.

We shall come back to Poisson noise in Chapter 7. Invariance of Poisson noise, in particular symmetric group and a characterization of Poisson noise will be discussed. We shall even speak of an infinite symmetric group in connection with its unitary representation.

Key point 3

Although we have discussed innovation of a stochastic process and a random field in the monograph[71], we shall discuss again in line with the idea *reduction* in Chapter 8. Weak sense innovation is also discussed briefly.

Part of the theory may be a rephrasement of the classical theory.

Chapter 9 contains variational calculus, which has also been discussed in the monograph[71]. We shall further discuss illustrative examples and operator fields defined by creation and annihilation operators. It is our hope that our approach to operator fields would contribute to quantum field theory.

The last chapter is devoted to four typical applications to quantum dynamics. They are the path integral with a development to the Chern–Simons action integral, infinite dimensional Dirichlet forms, the time operators and Euclidean field which is touched upon briefly.

The appendix contains some necessary formulas and notions in analysis and quick review of discrete parameter white noise, which will serve to good understanding of the basic ideas of white noise theory.

Having followed the chapters successively, the readers will see what is the white noise analysis and may answer the question on what are the advantages of this analysis practically. We repeat the characteristics and advantages which were briefly mentioned in the preface.

Chapter 2

Generalized white noise functionals

Before giving a definition of a stochastic process in general, we review two most important processes; Brownian motion and Poisson process. Then, we consider the meaning of "stochastice" in terms of Bernoulli's Ars Conjectandi and how to define a stochastic process. After that we can see that it is natural to use characteristic functional to define a stochastic process, and hence we are led to define generalized stochastic processes.

At this point, we just recall some examples of a basic stochastic process and briefly mention the idea of defining a stochastic process.

2.1 Brownian motion and Poisson process; elemental stochastic processes

White noise is a system of idealized elemental Gaussian random variables which are mutually independent and identically distributed. A realization of a white noise is given by the time derivative of a Brownian motion, denoted by $\dot{B}(t) = \dot{B}(t, \omega)$, $\omega \in \Omega(P)$, where $(\Omega, \mathbf{B}, P)$ is a probability space.

Some methods of defining a Brownian motion

There are many ways of defining a Brownian motion. Some of them are described in the following.

1) A classical and a standard way of defining a Brownian motion.

It is a traditional method, which gives us intuitive impression, being suggested by an irregular movement of a grain of pollen in the water.

Definition 2.1 A system of random variables $\mathbf{B} = \{B(t), t \in R\}$, is a

Brownian motion if the following three properties are satisfied:

(1) $\mathbf{B}$ is a Gaussian system with $E(B(t)) = 0$ for every t,
(2) $B(0) = 0$,
(3) $E[(B(t) - B(s))^2] = |t - s|$.

By definition, it follows that the covariance function $\Gamma(t,s)$ is

$$\Gamma(t,s) = E[(B(t)B(s)] = \frac{1}{2}(|t| + |s| - |t-s|). \tag{2.1.1}$$

The existence of a Brownian motion, as a Gaussian system, is obvious, since the function $\Gamma(t,s)$ is positive definite.

We now pause to discuss the term *additive process.* A stochastic process $X(t)$ is called an additive process if it has independent increments; namely for any $h > 0$, the increment $X(t+h) - X(t)$ is independent of the system $\{X(s), s \le t\}$.

It is easy to prove that Brownian motion is an additive process.

2) Standard method for the case where finite dimensional distribution is given.

For this case, we start with the covariance function $\Gamma(t,s)$ and mean zero. Finite dimensional Gaussian distribution is given by the mean vector and covariance matrix $\Gamma(t,s)$. Those finite dimensional Gaussian distributions are consistent. Thus we can appeal to the theory of Kolmogorov-Sinai that guarantees the existence and uniqueness of a probability distribution μ. Further it is known that the μ is defined on measurable space $(R^{[0,\infty)}, \mathcal{F})$ (for details see Hida-Hitsuda[73], Chap. 3), where $\mathcal{F}$ is the sigma field generated by cylinder subsets of $R^{[0,\infty)}$.

3) The method using the existence of a white noise.

It is the characteristic functional method, which is going to be explained in Section 2.3. Brownian motion is simply a definite integral of white noise. The existence of a Brownian motion is obvious.

Construction and approximation of Brownian motion

There are many well-known methods to construct a Brownian motion. Among them,

(1) Limit of normalized random walk by a suitable method of normalization. This method is well known, so it is omitted here.

(2) The Paley–Wiener's method of defining complex Brownian motion with parameter space $[0, 2\pi]$. This method is well known.

Take the parameter space to be $T = [-\pi, \pi]$. Suppose we take time derivative of a Brownian motion. Then, its covariance function $\gamma(t, s)$ is expressed formally in the form

$$\gamma(t, s) = \delta(t - s). \tag{2.1.2}$$

Thus we are given a generalized stochastic process, called *white noise.* Rigorous definition will be given in Section 2.3.

Now we are going to have a heuristic and somewhat formal observation on white noise, and with the help of the observation, we shall be back to a construction of a Brownian motion.

Delta function $\delta(t), t \in [-\pi, \pi]$ is expressed by Fourier series expansion of the form

$$\delta(t) = \frac{1}{2\pi} \sum_{-\infty}^{\infty} e^{int}. \tag{2.1.3}$$

This is the expansion of the covariance function of white noise, so we are (see (3) that follows) suggested to introduce a (generalized) stochastic process $X(t)$ by the formula:

$$X(t) = \frac{1}{2\pi} \sum_{-\infty}^{\infty} e^{int} X_n, \tag{2.1.4}$$

where $\{X_n, n \in Z\}$ is a system of independent random variables standard Gaussian in distribution.

Obviously the covariance function of $X(t)$ is equal to $\delta(t - s)$ given above.

Now, we note that $X(t)$ is a time derivative of a Brownian motion. So, we integrate $X(t)$ in the variable t to define a Brownian motion $Y(t), t \in T$;

$$Y(t) = \int_0^t X(s) ds$$

$$= \frac{1}{\sqrt{2\pi}} \sum_{n \in Z'} \frac{(e^{int} - 1)X_n}{in} + tX_0, \tag{2.1.5}$$

where $Z' = Z - \{0\}$.
As for the convergence of the above series, we can prove in the $L^2(\Omega, P)$ topology. Almost sure convergence can also be proved by using a trick (see Hida[40], Chapter 2).

(3) There is also a similar method that uses the so-called Karhunen-Loéve expansion (see Addenda A.2). In general such a method does not have much connection with what we are going to discuss. For instance, the so-called causal property is not so closely related. It is, however, worth to be reminded that the expansion is useful to construct a Brownian motion in connection with harmonic analysis, as will be seen later.

(4) P. Lévy's method of constructing a Brownian motion by using successive interpolation gives us many suggestions. See, for example Chapter 5. This method illustrates the property that Brownian motion is essentially infinite dimensional and is uniform in all directions. Actual form of this construction is now shown below.

Intuitively speaking, the Lévy construction of a Brownian motion is done by interpolation. It follows the following steps. Prepare standard Gaussian i.i.d. random variables $\{Y_n = Y_n(\omega)\}$ on a probability space $(\Omega, \mathbf{B}, P)$. Start with $\{X_1(t)\}$ given by

$$X_1(t) = tY_1. \tag{2.1.6}$$

The sequence of process $\{X_n(t), t \in [0,1]\}$ is formed by induction. Let T_n be the set of binary numbers $k/2^{n-1}, k = 0, 1, 2, \cdots, 2^{n-1}$, and set $T_0 = \cup_{n \ge 1} T_n$. Assume that $X_j(t) = X_j(t, \omega), j \le n$, are defined. Then, we set

$$X_{n+1}(t) = \begin{cases} X_n(t),\ t \in T_n, \\ \dfrac{X_n(t + 2^{-n}) + X_n(t - 2^{-n})}{2} + 2^{-\frac{n}{2}} Y_k,\ t \in T_{n+1} - T_n, \\ \qquad\qquad k = k(t) = 2^{n-1} + \frac{2^n t + 1}{2}, \\ (k + 1 - 2^n t) X_{n+1}(k 2^{-n}) + (2^n t - k) X_{n+1}((k+1)2^{-n}), \\ \qquad\qquad t \in [k2^{-n}, (k+1)2^{-n}], \end{cases} \tag{2.1.7}$$

where ω is omitted.

It is easy to see that the sequence $X_n(t), n \geq 1$, is consistent and that the uniform L^2-limit of the $X_n(t)$ exists. The limit is denoted by $\tilde{X}(t)$. It is proved that it has independent increments and

$$E(|\tilde{X}(t) - \tilde{X}(s)|^2) \leq |t - s|,$$
$$\Gamma(\tilde{X}(t), \tilde{X}(s)) = t \wedge s,$$

where Γ denotes the covariance.

Furthermore it can be proved that for almost all ω:

$$\lim_{n \to \infty} X_n(t, \omega) = X(t, \omega)$$

exists and is equal to $\tilde{X}(t)$, a.e. (P).

Summing up, we have proved

Proposition 2.1 *The process $X(t, \omega), t \in [0, 1]$, is a Brownian motion on the probability space $(\Omega, \mathbf{B}, P)$.*

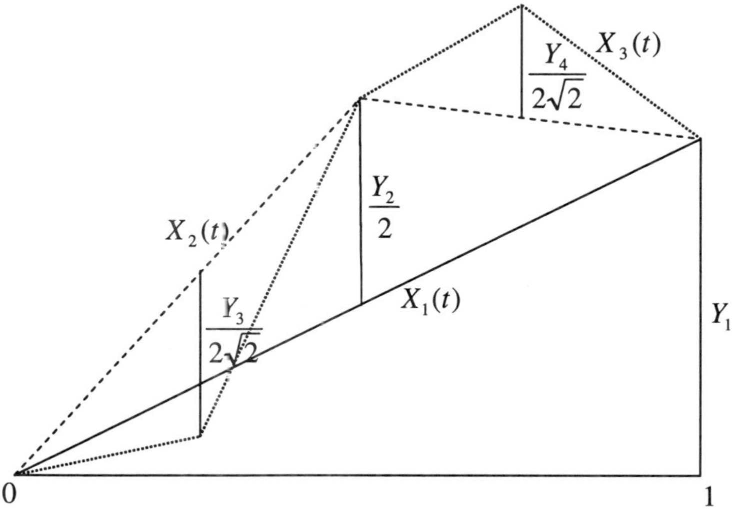

Fig. 2.1.

Each method mentioned above has its own interest. The method (1) may be thought of as a discrete (both in time and scale) approximation to a Brownian motion. As for method (2), one can find connection with harmonic analysis of white noise functionals. Since the method (4) is quite significant in many aspects, we will explain in detail, whenever we have the opportunity throughout this monograph. This shows not only computational interest, but also a description of hidden characteristic of Brownian motion. It is noted that this approximation to a Brownian motion is done uniformly in time.

Remark 2.1 *The method (4) shows that a system consisting of only countably many independent random variables can describe a Brownian motion, so that separability is guaranteed. However, if we take the time derivatives, the system* $\{\dot{X}(t), t \in [0,1]\}$ *spans, formally speaking, a linear space which has continuously many linearly independent vectors. We shall often meet such an interesting character of a Brownian motion (or of white noise) with two different sides. Although the second side is not quite rigorous.*

There is a significant application to white noise that comes from this method. White noise functionals can be approximated uniformly in time by using this approximation to Brownian motion. This idea implies many techniques, for example an irreducible unitary representation of a symmetric group can be obtained in the Hilbert space of nonlinear functionals of the square of white noise. There we need a suitable approximation, in fact the present method is just fitting. Another example arises when we have approximations where causality in time should always appear explicitly.

A Brownian motion is a Markov process, and it is known that almost all sample functions (or paths) $B(t,\omega)$ are continuous, indeed, uniformly continuous on any finite time interval, but they are not differentiable, (see e.g. Hida[45]).

Note. A multi-dimensional, say R^d-dimensional parameter Brownian motion is defined exactly the same as in Definition 2.1. It is called Lévy's Brownian motion. This process, actually a random field, will be discussed later.

Construction of a Poisson process.

A Poisson process denoted by $P(t) = P(t, \omega), t \geq 0, \omega \in \Omega$, is another basic stochastic process. It is defined in the classical manner as follows.

Definition 2.2 A stochastic process $P(t), t \geq 0$, is called a Poisson process if it satisfies the following three conditions:

1) $P(t)$ is additive; namely, for any $t > s \geq 0$, the increment $P(t) - P(s)$ is independent of the system $P(u), u \leq s$,
2) $P(0) = 0$,
3) $P(t) - P(s)$ is subject to a Poisson distribution with parameter λ.

The parameter λ is called the *intensity* of a Poisson process $P(t)$.

A Poisson process is an *additive process* by definition. It can be proved that almost all sample functions are monotone nondecreasing with discrete jumps of unit magnitude. We can form a version of a Poisson process such that almost all sample functions have discontinuity of the first kind and are right continuous. In what follows we always consider such a version.

It is better to recall the definition of a Lévy process.

Definition 2.3 A stochastic process $X(t) = X(t, \Omega), t \in [0, \infty), \omega \in \Omega$, is called a *Lévy process*, if it satisfies the conditions 1)– 4).

1) it is an additive process; for any $h > 0$ and any $t \geq 0$, the increment $X(t + h) - X(t)$ is independent of the system $\{X(s), s \leq t\}$.
2) $X(0) = 0$.
3) $X(t)$ is continuous in t in probability.
4) For almost all ω, $X(t, \omega)$ is a ruled function, i.e. it is right continuous in t and has left limit.

If, in particular, the probability distribution of the increment $X(t+h) - X(t), h > 0$ is independent of t, then the $X(t)$ is said to have *stationary increments* to be *temporary homogeneous.*

Obviously, both a Brownian motion and a Poisson process are temporary homogeneous Lévy processes.

There is a simple method of constructing a Poisson process. Let $X_n, n \geq 1$, be a sequence of independent, identically distributed random variables. Suppose the common distribution is exponential with parameter $\lambda > 0$; namely the density function is

$$f(x) = \begin{cases} e^{-\lambda x}, & x \geq 0 \\ 0, & x < 0. \end{cases} \tag{2.1.8}$$

Then a Poisson process $P(t)$ is defined in such a way that

$$P(t) = \begin{cases} 0, & t < X_1 \\ n, & \sum_{i=1}^{n-1} X_i \leq t < \sum_{i=1}^{n} X_i. \end{cases} \tag{2.1.9}$$

Setting

$$S(n) = \sum_1^n X_i,$$

we have a beautiful *duality*

$$P(t) > n \Leftrightarrow S(n) < t.$$

Incidentally $\sum_{i=1}^{\infty} X_i = \infty$ with probability 1, so the above definition makes sense. Indeed, we are given a version of a Poisson process with right continuous sample functions. Obviously the intensity is λ.

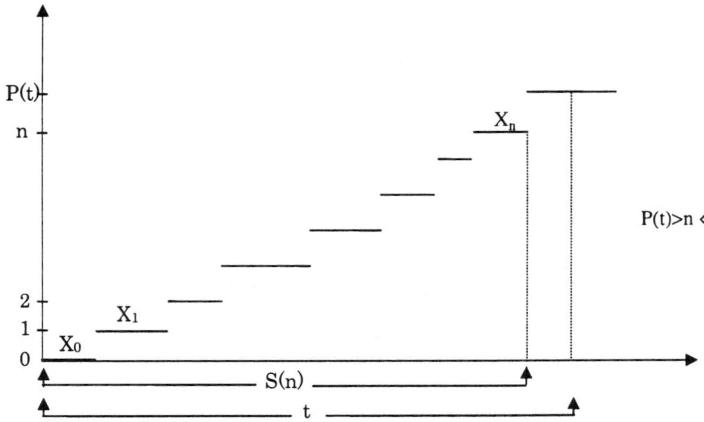

Fig. 2.2.

According to the Lévy decomposition of a Lévy process with stationary increments (see e.g. the books[36,101]), we may say that a Brownian motion and a Poisson process are atomic processes. So we give

Definition 2.4 Brownian motion as well as Poisson process are called *elemental additive processes.*

The time derivatives $\dot{B}(t)$ and $\dot{P}(t)$ are therefore generalized stochastic processes (in the sense of Gel'fand). In view of this, we can say that they are *elemental noises.*

They have significant different characteristics. A comparison of these typical noises will be discussed in the next section.

We have suggested many applications where the time development of random phenomena is discussed. It is, therefore, recommended to use the elemental noises explicitly.

2.2 Comparison between Brownian motion and Poisson process

Both a Brownian motion and a Poisson process are Lévy processes with stationary increments. They therefore have a lot of similar probabilistic properties. Moreover, as we have seen in the Lévy-Itô decomposition of a Lévy process, Lévy process consists of a Brownian motion and a compound Poisson process that involves many Poisson processes with various scales of jump. In addition, Brownian motion and Poisson process are elemental. In addition, Brownian motion and Poisson process can often be discussed in parallel, but it is not always so.

It is most important to discuss with emphasis on difference between the two. First a visualized appearance of sample function is noted. Almost all sample functions of a Brownian motion are continuous, although quite irregular. While those of a Poisson process is discontinuous; indeed it increases only by jumps of unit magnitude. This may be thought of as a simple difference; in fact, profound differences are there. When the Lévy-Itô decomposition is applied to a Lévy process, first we have to observe points where the process has discontinuity. Collecting all jumps, formally speaking, a compound Poisson process could be formed. The rest has to be a continuous component, so that it is proved to be Gaussian. Thus, the two are discriminated by sample function properties.

Another significant property of a sample function is seen from the behavior of fluctuating. In the case of a Brownian motion, almost every sample function behaves very irregularly in a special fashion of fluctuating. It is really a diffusion process and one of the realizations is seen in the expression of the diffusion equation, which is the same type as those of a heat equation. See e.g. Hida[40], Chapter 2. The equation is the second order parabolic partial differential equation. As for the Poisson case, sample functions are always non-decreasing without fluctuating left and right, so the associated differential equation is simply of the first order.

Now it seems better to make a short remark regarding the Gaussian and Poisson distributions. These two distributions enjoy the reproducing property. Namely, if two independent random variables X_1 and X_2 are subject to the distribution of the same type, so is the sum $X_1 + X_2$. It is an easy exercise to check that the reproducing property is satisfied by Gaussian and Poisson distributions, respectively.

Further remarkable difference between the two may be seen in terms of the noises obtained by taking the time derivatives of the respective processes, which are generalized stochastic process. For this purpose we discuss some background of the probability distributions of generalized processes that will be discussed in the next section. Some dissimilarities between two noises can be discussed in terms of characteristic functionals of (Gaussian) white noise and Poisson noise. The details will also be shown in Chapter 5, where we shall develop an infinite dimensional harmonic analysis that arises from rotation group and symmetric group, respectively.

We shall consider these noises again in Chapters 3 and 4.

2.3 The Bochner–Minlos theorem

We are now in a position to discuss a definition of noises, and more generally a method of characteristic functional to define stochastic processes as well as generalized stochastic processes. (Recall Section 1.3.)

To fix the idea, we start with the real Hilbert space $L^2(R)$. Take an increasing family $\{\|\cdot\|_n, n \geq 0\}$ of Hilbertian norms for which we assume that they are consistent. The 0-th norm $\|\cdot\|_0$ is tacitly understood to be the $L^2(R)$-norm $\|\cdot\|$. Let E_n be the Hilbert space, which is a dense subspace of $L^2(R)$ consisting of all members with finite $\|\cdot\|_n$-norm. Naturally E_n is topologized by the $\|\cdot\|_n$-norm. Let E_{-n} be the dual space of E_n with respect to the basic inner product in $L^2(R)$. Then, we have

$$\cdots \subset E_{n+1} \subset E_n \subset \cdots \subset L^2(R) \subset \cdots \subset E_{-n} \subset E_{-n-1} \subset \cdots.$$

All the mappings from left to right between subset sign are understood to be continuous injections.

Now assume that for any $m > 0$ there exists n $(> m)$ such that the injection

$$E_n \mapsto E_m$$

is of Hilbert-Schmidt type. Set

$$E = \bigcap E_n.$$

The topology of E is introduced in the usual manner. Then, the space E is called a *sigma-Hilbert nuclear* space or simply called a *nuclear* space. The dual space of E, which is to be the projective limit of the E_{-n}, is denoted by E^*.

Example 2.1 Set $\|\xi\|_p = \|A^p\xi\|$, $p \geq 1$, where

$$A = -\frac{d^2}{du^2} + u^2 + 1.$$

Then, obviously $\|\cdot\|_p$ is a Hilbertian norm. With the help of the Hilbertian norms $\|\xi\|_p, p \geq 0$, a nuclear space E is defined by the method explained above. The space E is the *Schwartz space* of test functions. It is a typical example of a nuclear space. The dual space E^* is the space of Schwartz distributions or the space of tempered distributions.

Let E^* be the dual space of E, i.e. the collection of all continuous linear functionals on E. A triple

$$E \subset L^2(R) \subset E^*$$

is called a *Gel'fand triple*. The space E is viewed as a space of test functions on R, while the space E^* is considered as the space of generalized functions.

The canonical bilinear form that connects E and E^* will be denoted as

$$\langle x, \xi \rangle, \quad x \in E^*, \xi \in E.$$

Given such a space E^*, we can form a measurable space $(E^*, \mathcal{B})$, where $\mathcal{B}$ is a sigma-field generated by the cylinder subsets of E^*.

A functional $C(\xi)$ on a nuclear space E is called a *characteristic functional* if it satisfies the following conditions :

1) $C(\xi)$ continuous on E,
2) $C(0) = 1$, and
3) $C(\xi)$ is positive definite; namely, for any n, and for any choices of $z_j \in \mathbf{C}$ and ξ_j, $1 \le j \le n$,

$$\sum_{1 \le j,k \le n} z_j \bar{z}_k C(\xi_j - \xi_k) \ge 0.$$

A characteristic functional is a generalization of a characteristic function of a probability distribution on R^d. In fact, we can prove the following theorem.

Theorem 2.1 *(Bochner–Minlos) Let $C(\xi)$ be a characteristic functional. Then there exists a probability measure μ on the measurable space $(E^*, \mathcal{B})$ such that*

$$C(\xi) = \int_{E^*} \exp[i\langle x, \xi \rangle] d\mu(x).$$

Such a measure is unique.

For proof see e.g. the monograph[40] Chapter 3.

Definition 2.5 A measure space $(E^*, \mathcal{B}, \mu)$ or simply written $X = (E^*, \mu)$ is called a *generalized stochastic process* with the characteristic functional $C(\xi)$.

There is a variation of this theorem.

Theorem 2.2 *If, in particular, the given $C(\xi)$ is continuous on $E_m, m \ge 0$, then the measure μ is supported by E_n^*, where $n > m$ and the injection $E_n \to E_m$ is Hilbert–Schmidt type.*

Example 2.2 Take $C(\xi) = \exp[-\frac{1}{2}\|\xi\|^2]$. Then, it is easy to see that $C(\xi)$ is a characteristic functional. The measure μ defined by $C(\xi)$ is the *white noise measure.* We can check that for the time derivative $\dot{B}(t)$ of a Brownian motion $B(t), t \in R$, the characteristic functional is

$$E(\exp[i\langle \dot{B}, \xi\rangle]) = \exp[-\frac{1}{2}\|\xi\|^2].$$

Since this characteristic functional is continuous in the topology determined by the norm $\|\cdot\|$, the white noise measure μ is supported by $E_{-1} = E_1^*$ from Theorem 2.1.

Definition 2.6 The measure space (E^*, μ) is called a *white noise.*

We know that μ-almost all $x \in E_1^*$ are viewed as sample functions of $\dot{B}(t)$. In view of this, $\dot{B}(t)$ is also called white noise.

Remark 2.2 *A generalized stochastic process is called stationary if the associated measure μ is invariant under the time shift. In this case, we can speak of the spectrum. If the spectral measure is flat, then as in Optics we say that it is white (colorless). To avoid this term, the above example of a generalized stochastic process may be called a Gaussian white noise, if one is afraid of confusion.*

Example 2.3 For the time derivative $\dot{P}(t)$ of a Poisson process $P(t)$ we have

$$E(\exp[i\langle \dot{P}, \xi\rangle]) = \exp[\lambda \int (e^{i\xi(t)} - 1)dt],$$

where $\lambda > 0$ denotes the intensity. The associated probability measure on E^* is denoted by μ_P.

This characteristic functional is continuous in the $\|\cdot\|_1$-norm, so that the measure μ_P is supported by E_{-2}.

There is a short remark. In the case where the time parameter t is restricted to a finite interval, say $[0,1]$, the characteristic functional is continuous in $L^2([0,1])$-norm. Hence μ_P is supported by the space E_{-1}, which is a subspace of $L^2([0,1])$.

2.4 Observation of white noise through the Lévy's construction of Brownian motion

Recall Lévy's construction of Brownian motion discussed in Section 2.1.

A notion to be reminded is that a white noise is a probability measure space (E^*, μ) determined by the characteristic functional

$$C(\xi) = \exp[-\frac{1}{2}\|\xi\|^2],$$

where E is a nuclear subspace of $L^2([0,1])$.

Now take a complete orthonormal system $\{\xi_n\}$, such that every ξ_n is in E. Then, it is easy to see that the system $\{\langle x, \xi_n \rangle\}$ forms an i.i.d. (independent identically distributed) random variables, each of which is subject to the standard Gaussian distribution $N(0,1)$. For simplicity we set

$$Y_n (= Y_n(x)) \equiv \langle x, \xi_n \rangle.$$

We now come to the Lévy construction of a Brownian motion by interpolation. Let us review it. In (2.1.6), $\{X_1(t)\}$ is defined and in (2.1.7), the sequence of processes $\{X_n(t), t \in [0,1]\}$ is formed by induction. Note that the T_n is the set of binary numbers $\frac{k}{2^{n-1}}, k = 0, 1, 2, \cdots, 2^{n-1}$, and set $T_0 = \cup_{n \geq 1} T_n$.

Thus, except the binary points, we can take time derivatives $\dot{X}_n(t)$.

It is easy to see the following.

Proposition 2.2 *i) The sequence $\dot{X}_n(t), n \geq 1$, is consistent, and the limit of the $\dot{X}_n(t)$ exists as a generalized stochastic process, the limit is denoted by $\dot{X}(t)$.*

ii) The $\dot{X}(t)$ is a (Gaussian) white noise.

Proof. We compute the characteristic functionals by successive approximation by evaluating conditional expectations.

■

2.5 Spaces (L^2), F and $\mathcal{F}$ arising from white noise

We now take a white noise measure μ introduced on the measurable space $(E^*, \mathcal{B})$. In the usual manner, a complex Hilbert space $(L^2) = L^2(E^*, \mathcal{B}, \mu)$ is formed. It is a collection of complex valued functionals of white noise with finite variance. *The White Noise Analysis* should start with this Hilbert space (L^2).

Also, we wish to utilize the traditional method of analysis starting from the analysis of elementary functionals. Having done so, we propose to provide the space of generalized white noise functionals at the most basic level. This will be discussed in the next section.

Method of characteristic functional

Take the characteristic functional $C(\xi) = \exp[-\frac{1}{2}\|\xi\|^2]$, $\xi \in E$, of $\dot{B}(t), t \in R$. Since it is positive definite, there exists a *Reproducing Kernel Hilbert Space* (RKHS) $\mathbf{F}$ with reproducing kernel $C(\xi - \eta)$, $(\xi, \eta) \in E \times E$. See Addenda A.3 for RKHS.

The reproducing kernel admits a power series expansion:

$$C(\xi - \eta) = \sum_{n=0}^{\infty} C_n(\xi - \eta),$$

where

$$C_n(\xi - \eta) = C(\xi)\frac{(\xi, \eta)^n}{n!}C(\eta).$$

It is easy to show that for any n, $C_n(\xi - \eta)$, $(\xi, \eta) \in E \times E$, is positive definite, and it generates a subspace $\mathbf{F}_n$ of $\mathbf{F}$ with reproducing kernel C_n. We claim

Proposition 2.3 *i) The reproducing kernel holds the following property:*

$$(C_n(\cdot - \eta), C_m(\cdot - \zeta))_{\mathbf{F}} = 0, \quad n \neq m. \tag{2.5.1}$$

ii) The space $\mathbf{F}$ *admits a direct sum decomposition:*

$$\mathbf{F} = \bigoplus_{n=0}^{\infty} \mathbf{F}_n. \tag{2.5.2}$$

Proof. i) The reproducing property of the kernel shows

$$(C(\cdot - t\eta), C(\cdot - s\zeta))_{\mathbf{F}} = C(s\zeta - t\eta).$$

By using the expansion of $C(\zeta - \eta)$, we have

$$\left(C(\cdot) \sum_0^\infty \frac{(\cdot, t\eta)^n}{n!} C(t\eta), C(\cdot) \sum_0^\infty \frac{(\cdot, s\zeta)^m}{m!} C(s\zeta) \right)_{\mathbf{F}} = C(s\zeta - t\eta).$$

Hence, we have

$$\left(C(\cdot) \sum_0^\infty \frac{(\cdot, \eta)^n t^n}{n!} C(\eta), C(\cdot) \sum_0^\infty \frac{(\cdot, \zeta)^m s^m}{m!} C(\zeta) \right)_{\mathbf{F}} = \frac{C(s\zeta - t\eta)C(\eta)C(\zeta)}{C(s\zeta)C(t\eta)}.$$

By using the expression of $C(\xi)$, the right–hand side becomes

$$\exp[st(\eta, \zeta)]C(\eta)C(\zeta).$$

Apply the operator $\frac{\partial^{n+m}}{\partial t^n \partial s^m}|_{t=s=0}$ to prove equations (2.5.1) and (2.5.2).

ii) We therefore have for $n \neq m$

$$\mathbf{F}_n \perp \mathbf{F}_m.$$

Recall the expansion of $C(\xi - \eta)$. Then, we finally prove the direct sum expansion of $\mathbf{F}$.

■

The $\mathcal{T}$-transform and the S-transform

We now discuss another type of transformations acting on the spaces of white noise functionals $\varphi(x)$'s in (L^2). cf. Addenda A.1.

The first attempt to get a good representation of φ was an infinite dimensional analogue of the Fourier transform. There were many plausible reasons to choose the $\mathcal{T}$-transform (see the references[61,40,41]) defined by

$$(\mathcal{T}\varphi)(\xi) = \int_{E^*} \exp[i\langle x, \xi\rangle]\varphi(x)d\mu(x), \ \ \xi \in E.$$

It plays some roles as we have expected; for example, the image $(\mathcal{T}\varphi)(\xi)$ is usually a nice (in analytic sense) functional of C^∞-function ξ, so that it is ready to be analyzed by functional analysis, instead of original nonlinear function $\varphi(x)$ of a generalized function x. Also it plays partly roles of the Fourier transform as can be guessed by its formula.

Some years later, I. Kubo and S. Takenaka[89] introduced the so-called S-transform, a nice modification of the $\mathcal{T}$-transform of (L^2)-functionals. For $\varphi(x) \in (L^2)$, the $(S\varphi)(\xi)$ is given by

$$(S\varphi)(\xi) = C(\xi) \int_{E^*} \exp[\langle x, \xi \rangle]\varphi(x)d\mu(x). \tag{2.5.3}$$

It may also be defined in such a manner that

$$(S\varphi)(\xi) = \int_{E^*} \varphi(x+\xi)d\mu(x).$$

To show this equality, we should note that the white noise measure μ is translation *quasi-invariant*. In fact, if x is shifted as much as $f \in L^2(R)$, then we can show that

$$d\mu(x+f) \approx d\mu(x), \quad f \in L^2(R),$$

where $\approx$ means the equivalence. In addition, we should note that the explicit form of the Radon-Nikodym derivative is given by

$$\frac{d\mu(x+f)}{d\mu(x)} = \exp[-\langle x, \xi \rangle - \frac{1}{2}\|f\|^2].$$

For proof, see Hida[40] Chapter 5.

■

Warning The functional $S(\varphi)(\xi)$ is often called the *U-functional* associated with φ because of Theorem 2.4. We, therefore, use this term when $U(\xi)$ is known to be the S-transform of a white noise functional.

Remark 2.3 *The S-transform may be thought of as an analogue of the Laplace transform applied to functionals on R^n. Indeed, the transform satisfies many analogous properties to those of the Laplace transform of functions defined on R^d. However, in what follows, one can see gradually many favorable properties when the analysis on (L^2) is proceeding. It is quite important to note that the S-transform is entirely different, in nature and in action, from the classical transformations that give homeomorphic automorphisms of Hilbert space. In our case, under the S-transform, we are* *given a reproducing kernel Hilbert space which provides better analytic expressions of generalized white noise functionals. For example, renormalized polynomials in* $\dot{B}(t)$*'s are carried to those in* $\xi(t), \xi \in E$*, Delta functions of* $B(t)$ *can also be transformed to a functional that is easy to handle.*

Another significant remark is that the S*-transform of* $\varphi(x)$ *is obtained by taking the expectation with respect to Gaussian measure, i.e. by phase average, where* (E^*, μ) *is viewed as a phase. While, it is proved, in Chapter 5, that the shift (the time shift) is a one-parameter subgroup of the infinite dimensional rotation group and is ergodic. This implies, roughly speaking, that the phase average is in agreement with the time average, Hence, the* S*-transform of* φ *is equal to the time average of* $\exp[x, \xi]\varphi(x)$ *up to the universal factor* $C(\xi)$*. In statistical application, phase (or ensemble) average is not easy to obtain, but time average can be computed by an observation of a time series. This is again advertized as an advantage of the* S*-transform.*

We now give some illustration of the S-transform.

1) First, the S-transform carries generalized white noise functionals $\varphi(x)$ to a visualized functionals $(S\varphi)(\xi)$ which have analytic properties. In general $\varphi(x)$ is nonlinear functional of a generalized function x, so that it has complex expression. By applying S-transform, $\varphi(x)$ becomes easy to observe and ready to be analyzed. In this sense, S gives a representation of $\varphi(x)$.

2) Second, the image $\mathcal{F}$ of (L^2) under S is topologized to be a *reproducing kernel Hilbert space* (abbr. **RKHS**) in order to establish an isomorphism between $\mathcal{F}$ and (L^2). As a result, different structures of those spaces provide more techniques for the analysis. The space $\mathcal{F}$ involves mostly functionals which are familiar in classical functional analysis, so that there are many handy tools around the house, which means that they are ready to be analyzed.

Note that the isomorphism is obviously not an automorphism of (L^2).

3) Third, as we shall see in the next section, S-transform extends to a mapping from the space of generalized white noise functionals and good representations of those functionals, too. This fact can be realized because $\exp[\langle x, \xi\rangle]$ is a test functional.

For instance

$$S(: x(t)^2 :)(\xi) = \xi(t)^2.$$

Here $: x(t)^2 :$ denotes the Wick product of degree 2. General Wick product will be given later in Example 2.6 in terms of $\dot{B}(t)$. At present, $: x(t)^2 :$ is understood as $x(t)^2 - \frac{1}{dt}$.

4) The $\mathcal{T}$-transform was defined in the spirit of the Fourier transform. Both $\mathcal{T}$ and S transforms are useful and each one plays its own role.

5) Since the S-transform is obtained by the integration with respect to μ, which means phase (ensemble) average of functions of samples (paths or trajectories of a dynamical system). We know that the calculation of S-transform can be obtained by applying the *ergodic* theory instead of integration with respect to μ, since phase average in question is in agreement with time average, this means, in particular, that the S-transform can be obtained from single sample function by taking the time average. This property is quite useful in applications to physics.

The facts illustrated above are now going to be rigorously proved and concrete results will be established as follows.

As is briefly mentioned in 2) above we have a space $\mathcal{F}$:

$$\mathcal{F} = \{S\varphi; \varphi \in (L^2)\} \tag{2.5.4}$$

forms a vector space over $\mathbf{C}$ as is easily proved.

Proposition 2.4 *i)* $(S\varphi)(\xi)$ *is continuous in* $\xi \in E$.

ii) The S*-transform gives a homeomorphism*

$$(L^2) \sim \mathcal{F}. \tag{2.5.5}$$

Proof. i) The continuity of $(S\varphi)(\xi)$ in ξ comes from the following fact.

$$\int e^{\langle x,\xi\rangle}\varphi(x)d\mu(x) - \int e^{\langle x,\eta\rangle}\varphi(x)d\mu(x) = \int (e^{\langle x,\xi\rangle} - e^{\langle x,\eta\rangle})\varphi(x)d\mu(x). \tag{2.5.6}$$

The integrand is bounded by

$$\left|(e^{\langle x,\xi\rangle} - e^{\langle x,\eta\rangle})\varphi(x)\right|$$

which is integrable and

$$e^{\langle x,\xi\rangle} - e^{\langle x,\eta\rangle} \to 0 \text{ as } \xi \to \eta \text{ in } E.$$

Hence we have proved the continuity of $(S\varphi)(\xi)$ in ξ.

ii) First we show that S is bijective:

$$(L^2) \mapsto \mathcal{F},$$

proving the continuity of $(S\varphi)(\xi)$ in ξ.

This property is proved by the fact that the algebra generated by

$$\{\exp[a\langle x,\xi\rangle],\ a \in R,\ \xi \in E\}$$

is dense in (L^2).

The continuity of the bijection above can be proved in the similar manner as i).

We shall now establish the relationship between the **RKHS F** with reproducing kernel $C(\xi-\eta), C(\xi)$ being the characteristic functional of white noise, and $\mathcal{F}$.

The subspace $\mathbf{F}_n$ of RKHS $\mathbf{F}$ has an element $C(\xi)(\xi,\eta)C(\eta)$, which plays the role of the reproducing kernel. Namely, we have

$$(C(\cdot)(\cdot,\eta)^n C(\eta), C(\cdot)(\cdot,\zeta)C(\zeta))_{\mathbf{H}_n} = C(\zeta)(\zeta,\eta)^n C(\eta).$$

Ignoring the functional $C(\xi)$ that is the universal factor, we may be allowed to write simply

$$((\cdot,\eta)^n, (\cdot,\zeta)) = (\zeta,\eta)^n.$$

Further, we may write

$$(\cdot,\eta)^n \sim \eta^{n\otimes} \in L^2(R)^{n\otimes}.$$

Finally, we have established the isomorphism

$$\mathbf{F}_n \cong \widehat{L^2(R^n)}, \tag{2.5.7}$$

where $\widehat{L^2(R^n)}$ denotes the subspace of $L^2(R^n)$ involving all symmetric functions.

On the other hand, the S-transform of a Fourier-Hermite polynomial $H_n(\langle x, \eta\rangle/\sqrt{2})$ with $\|\eta\| = 1$, is $\langle \eta^{n\otimes}, \xi^{n\otimes}\rangle$, and this relation extends to the isomorphism between the subspace H_n of (L^2) where

$$H_n = \text{subspace of } (L^2) \text{ spanned by } H_n(\langle x, \eta\rangle), \eta \in E. \qquad (2.5.8)$$

The subspace H_n of (L^2) is the space of the so-called *homogeneous chaos* or the space of the *multiple Wiener integrals*, in Itô's sense, of degree n.

Let $\mathcal{F}_n$ be defined by the restriction of the isomorphism given by (2.5.5) to H_n.

With this notation and with the summary of what we have discussed, we have now established the theorem below.

Theorem 2.3 *The following three direct sum decompositions are equivalent.*

$$(L^2) = \bigoplus_{n=0}^{\infty} H_n,$$

$$\mathcal{F} = \bigoplus_{n=0}^{\infty} \mathcal{F}_n,$$

$$\mathbf{F} = \bigoplus_{n=0}^{\infty} \mathbf{F}_n.$$

Each space with the direct sum decompositions as above may simply be called the *Fock space.*

At this stage some interpretation on Fock space is necessary. Here we do not give a mathematical definition of Fock space. We will, however, tacitly understand that it is a Hilbert space H which admits a direct sum decomposition into mutually orthogonal subspaces H_n of the form

$$H = \bigoplus_{n=0}^{\infty} H_n,$$

satisfying the following conditions 1) and 2):

1) Each H_n is a graded subspace, to which degree n is associated.

2) There exists a mapping ∂_n which is a continuous surjection

$$\partial_n \ : \ H_n \ \rightarrow \ H_{n-1}.$$

The operator ∂_n is to be an annihilation operator. In fact, we can define such an operator in the following manner.

In the present case H is taken to be (L^2). Then, take a complete orthonormal system $\{\xi_n, n \geq 1\}$. If $\varphi(x)$ in (L^2) is expressed in the form

$$\varphi(x) = f(\langle x, \xi_1\rangle, \langle x, \xi_2\rangle, \cdots, \langle x, \xi_n\rangle)$$

for some smooth function $f(t_1, t_2, \cdots, t_n)$ on R^n, then

$$\partial_k \varphi(x) = \frac{\partial}{\partial t_k} f(t_1, t_2, \cdots, t_n)|_{t_j=<x,\xi_j>}. \tag{2.5.9}$$

So far as $\{\xi_n\}$ is fixed, the operator ∂_k is well defined for smooth *tame functionals* satisfying the above assumptions; namely, $\partial_k \varphi$ is uniquely determined regardless of the representation by a function f.

The class of smooth tame functionals is dense in (L^2), so that we claim that the domain $\mathcal{D}(\partial_k)$ is dense in (L^2).

By the requirement 2), the adjoint operator of ∂_n, denoted by ∂_n^*, is defined in such a way that

$$(\partial_n f, g) = (f, \partial_n^* g), \ f \in H_n, \ g \in H_{n-1},$$

where $(\cdot, \cdot)$ is the inner product in $H = (L^2)$. It can be proved that the operator ∂_n^* is a continuous mapping from H_{n-1} to H_n.

It is easy to see that ∂^* with the property

$$(\partial f, g) = (f, \partial^* g), \ f, g \in H,$$

can be defined.

The reader is led to have a bird's-eye view of what have been discussed in this section by observing the following triangle.

RKHS

$$\mathbf{F}$$

$$\swarrow \qquad\qquad \searrow$$

$$(L^2) \qquad \longleftrightarrow \qquad \mathcal{F}$$

Diagram 2.1

2.6 Generalized white noise functionals

Generalized white noise functional is one of the most significant subjects of our white noise theory. Indeed, the study of generalized white noise functionals is the foundation of the advanced theory.

We have started with Brownian motion and its functionals in the last subsection, and the theory has been established to some extent. For further developments, we have decided to start the theory more systematically, that is, in line with Reduction, Synthesis and Analysis as was explained in Chapter 1. As a result, we can answer the question on how widely the functionals are introduced and what should be discussed. We are naturally led to take white noise as a variable of functionals in question and to introduce generalized white noise functionals. Operators acting on the spaces that have been determined will be discussed in the next section.

We are now in a position to consider what kind of problems can be and should be dealt with. We will see the problems which are related with generalized white noise functionals. They are now in order.

(1) The notation $\dot{B}(t)$ is just a formal expression, although we can understand its intuitive meaning as is well-known. It cannot be a random variable in the ordinary sense, therefore its probability distribution can not be defined. A first correct understanding is given with the help of

characteristic functional which is of the form $C(\xi) = \exp[-\frac{1}{2}\int \xi(t)^2 dt]$. It is computed in such a manner that $E(\exp[i\langle \dot{B}, \xi\rangle])$, where we know

$$\langle \dot{B}, \xi\rangle = \dot{B}(\xi) = -\langle B, \xi'\rangle. \tag{2.6.1}$$

Hence, we are given a generalized stochastic process with independent values at every moment t in the sense of Gel'fand[28]. Note that the $\dot{B}(\xi)$ has meaning, but not $\dot{B}(t)$.

(2) We remind the applications to Wiener's theory on nonlinear networks (see Wiener[166]) and nonlinear prediction theory (see the literature[114]). There random source occurs at instant t and it plays a significant role. We therefore wish to deal with $\dot{B}(t)$ itself.

(3) It would be fine if stochastic differential equations can be dealt with as if it were ordinary differential equations, although the variable is taken to be $\dot{B}(t)$.

(4) Problems in quantum dynamics. There is a proposed approach to Tomonaga-Schwinger equations, where we use operators that come from differential operators in $\dot{B}(t)$. Analysis of random fields and quantum fields is a direct application. We shall discuss Feynman's path integrals.

(5) The infinite dimensional rotation is used in connection with white noise measure μ which is the probability distribution of white noise.

The effective use of generalized white noise functionals is fundamental in our analysis as we shall see in many places of the present monograph. In addition, we say that our analysis has been quite naturally set up exactly in line with the traditional method of the analysis and with new ideas on how to treat random functions that are in question. As a result, generalized white noise functionals play important roles as well as in applications.

The idea of introducing generalized white noise functionals arose many years ago, around 1970 with the above facts (1)–(5) in mind.

It was until 1975, the idea of introducing such functionals was presented in the introduction of the Carleton Mathematical Notes, and we actually proposed to establish the analysis of them. There was given a guiding expression of the form

$$\varphi(\dot{B}(t); t \in T),$$

although a formal expression, where T is an interval and $\dot{B}(t)$'s are variables of φ. The φ is, of course, nonlinear in general. The $\dot{B}(t)$ is nothing but a realization of white noise.

Again we remind that $\dot{B}(t), t \in R$, has been so far understood as a generalized stochastic process in the sense of Gel'fand, namely it is parametrized by ξ that runs through a nuclear space E. For our purpose this is not satisfied. In fact, the parameter should be $t \in R$. In other words, single $\dot{B}(t)$ should have, as it were, an *identity*. We now establish a definition of $\dot{B}(t)$.

One can get the smeared variable

$$\langle \dot{B}, \xi \rangle = \int \dot{B}(t)\xi(t)dt = -\int B(t)\xi'(t)dt, \tag{2.6.2}$$

in the classical sense. Note that the time variable t disappears. This is a crucial disadvantage, when we discuss time development of random phenomena explicitly in terms of t. In reality, if $\dot{B}(t)$ denotes a fluctuation or a random input signal to a channel that happens at instant t, then the evolutional phenomena interfered with by the fluctuation should be expressed in terms of $\dot{B}(t)$ itself without being smeared.

Coming back to $\langle \dot{B}, \xi \rangle$ which we shall denote simply by $\dot{B}(\xi)$. To fix the idea, the parameter space is taken to be $I = [0,1]$. We see that the covariance is

$$E(\dot{B}(\xi)\dot{B}(\eta)) = \int_0^1 \xi(t)\eta(t)dt.$$

This equality enables us to extend $\dot{B}(\xi)$ to $\dot{B}(f), f \in L^2(I)$ which is a Gaussian random variable with probability distribution $N(0, \|f\|^2)$. This is a stochastic bilinear form defined almost everywhere (μ). This fact implies that there exists an isomorphism between the subspace H_1 (a member of the Fock space, where the time parameter set is I) and the Hilbert space $L^2(I)$ as was discussed before:

$$H_1 \cong L^2(I). \tag{2.6.3}$$

The ideal random variable $\dot{B}(t)$ corresponds to the case where the test function ξ is replaced by the delta function δ_t which is in the Sobolev space of order -1, denoted by $K^{(-1)}(I)$, which is the dual space of $K^{(1)}(I)$ of order 1. We have the inclusions

$$K^{(1)}(I) \subset L^2(I) \subset K^{(-1)}(I),$$

where the mapping from left to right is continuous injection of Hilbert Schmidt type.

First we restrict the isomorphism (2.6.4) to

$$K^{(1)}(I) \cong H_1^{(1)}.$$

Actually $H_1^{(1)}$ is defined so as to hold the above isomorphism. The dual space $H_1^{(-1)}$ of $H_1^{(1)}$ can be defined and we have

$$H_1^{(1)} \subset H_1 \subset H_1^{(-1)} \tag{2.6.4}$$

in parallel with

$$K^{(1)}(I) \subset L^2(I) \subset K^{(-1)}(I).$$

A member of $H_1^{(-1)}$ may be written as $\dot{B}(g), g \in H^{(-1)}(I)$ which is an extension of a stochastic bilinear form. Now we are happy to take g to be a delta function δ_t to have $\dot{B}(\delta_t) = \dot{B}(t)$.

¿From now on, $\dot{B}(t)$ is defined in the sense established above. Members in $H_1^{(-1)}$ are called linear *generalized white noise functionals.* We may say that identity of $\dot{B}(t)$ is given as a member of $H_1^{(-1)}$.

Remark 2.4 *If we apply the variation of the Bochner-Minlos Theorem 2.2, we see that the characteristic functional* $C(\xi) = \exp[-\frac{1}{2}\|\xi\|^2]$ *uniquely determines the white noise measure on* $K^{(-1)}(I)$*, so that we may say that almost all sample functions of* $\dot{B}(t), t \in I$*, are in* $K^{(-1)}(I)$.

Having been given the system of variables $\{\dot{B}(t)\}$ we can establish a class of acceptable and meaningful functions, actually functionals, following the standard method in analysis. First we have to define the most elementary and basic functions, namely polynomials in $\dot{B}(t)$'s. For example, we first give $(\dot{B}(t))^n$, in particular $(\frac{dB}{dt})^2$.

Now, remind the simplest case of the Itô formula $(dB(t))^2 = dt$. But this does not allow us to have $(\frac{dB}{dt})^2 = \frac{1}{dt}$, which has no correct meaning. However, $(dB(t))^2$ is random, although it is infinitesimal. To define the square of $\dot{B}(t)$, we need some modification to overcome the difficulties above. Our idea is to apply the *renormalization* to the centered random quantity $(dB(t))^2$. The trick is as follows. First subtract off the expectation to have $(dB(t))^2 - dt$ and then by multiplying as much as $(\frac{1}{dt})^2$. As a formal expression we must have a difference $\dot{B}(t)^2 - \frac{1}{dt}$, however each term of this difference has no meaning. We therefore approximate by $(\frac{\Delta B}{\Delta})^2 - \frac{1}{\Delta}$, Δ being an interval involving t, and let $\Delta \to \{t\}$. Then, there exists a limit in $H_2^{(-2)}$. Needless to say, the limit is taken in the topology of $H_2^{(-2)}$. Thus, a fully sensible meaning can be assigned to the limit which is allowed to write $\dot{B}(t)^2 - \frac{1}{dt}$, noting that this difference should be considered as a single entity.

To define a big space that includes renormalized $\dot{B}(t)^2$ and more general renormalized polynomials, we may think of a generalization of what we did for $\dot{B}(t)$. For general functionals of $\dot{B}(t)$'s, generalization is too complicated. We can, however, provide an efficient tool for this purpose. It is the S-transform rephrased in the form

$$S(\varphi(B))(\xi) = \exp[-\frac{1}{2}\|\xi\|^2]E[e^{<\dot{B},\xi>}\varphi(\dot{B})].$$

If we apply the S-transform, we are given acceptable quantity $\xi(t)^2$ rigorously. It is associated with the renormalized square of $\dot{B}(t)$; we denote it by $:\dot{B}(t)^2:$. Rigorous definition is seen later in Example 2.4.

The above observation suggests that renormalized polynomials in $\dot{B}(t)$'s can be made to be generalized functionals of white noise as is seen by applying the S-transform.

There is an important remark. The parameter set has been taken so far to be $I = [0, 1]$ for convenience. We can easily extend the results to the case where parameter set is R with minor modification. So, we use the same notations like H_1 and $H_1^{(-1)}$, while $K^{(-1)}(I)$ will be changed to $K^{(-1)}(R)$ and so forth.

We are now ready to define a class of generalized white noise functionals in the general setup. There are two typical ways of introducing generalized white noise functionals rigorously. One is to use the Sobolev space structure and the other is an (infinite dimensional) analogue of the Schwartz space of the tempered distributions on R^d.

A. Use of the Sobolev space structure

We are now ready to introduce generalized white noise functionals in a general setup, by using Sobolev spaces. (See, e.g. Lions and Magenes[110].) Note the S-transform. If it is restricted to H_1, that is the space of degree 1, is mapped to $L^2(R)$. Generally, we know that there is a mapping π_n that defines an isomorphism

$$\pi_n : \quad H_n \cong \widehat{L^2(R^n)}, \tag{2.6.5}$$

where H_n is the space of homogeneous chaos of degree n and where $\widehat{L^2(R^n)}$ is the subspace of $L^2(R^n)$ consisting of all symmetric functions. More precisely, for $\varphi \in H_n$ the following equality holds:

$$\|\varphi\| = \sqrt{n!}\|\pi_n\varphi\|_{L^2(R^n)}.$$

For convenience, the symbol π_n will be used to denote the mapping from H_n to $\widehat{L^2(R^n)}$.

Our basic idea is to have a reasonable extension of the space H_n corresponding to a reasonable extension of $\widehat{L^2(R^n)}$.

There is hope to have such an extension of H_n. In fact, this is possible in a quite natural manner. One of the significant reasons is that H_n is an infinite dimensional vector space. If the basic space were finite dimensional, one must find some unconventional reason why it can naturally be extended. The space H_n, accordingly $\widehat{L^2(R^n)}$, can find its own course to be extended to a space of certain generalized functions. Moreover, there are good examples to be taken care of; namely polynomials in $\dot{B}(t)$'s are not in (L^2), but should be included in some extended spaces. Suitable additive renormalization acting on those polynomials leads us to introduce generalized white noise functionals that correspond, by the above isomorphism, to generalized functions on R^n.

For example, the renormalized $\dot{B}(t)^n$, denoted by $:\dot{B}(t)^n:$ is a generalized white noise functional that should belong to an enlarged space of H_n. Note that the most important idealized random variable $\dot{B}(t)$ itself is a generalized white noise functional that corresponds to the delta function $\delta_t(\cdot)$.

A mathematically rigorous construction of the proposed space, obtained with the help of the Sobolev spaces, is as follows.

Denote by $K^m(R^n), m > 0$, the Sobolev space over R^n of order m. The subspace invoking all symmetric functions of $K^m(R^n)$ will be denoted by $\widehat{K^m(R^n)}$. The space $\widehat{K^{-m}(R^n)}$ is the dual space of $\widehat{K^m(R^n)}$ with respect to the topology being equipped with $L^2(R^n)$. Thus we have another triple for every $m > 0$:

$$\widehat{K^m(R^n)} \subset \widehat{L^2(R^n)} \subset \widehat{K^{-m}(R^n)},$$

where both inclusions are continuous injections.

Our favourite choice of the degree m of the Sobolev space is taken to be $(n+1)/2$. For one thing, members in $K^{(n+1)/2}(R^n)$ are continuous and the restriction to an $(n-1)$-dimensional hyperplane belongs to the space $\widehat{K^{n/2}(R^{n-1})}$, namely the degree decreases by $1/2$ when the restriction is made to a subspace of one dimension lower. This comes from the trace theorem in the theory of the Sobolev spaces. Moreover, if the trace of $K^{(n+1)/2}(R^n)$-function is taken by the integral on the diagonal, then the degree decreases by one. These properties are convenient when the analysis on H_n is carried out, by taking the isomorphism between H_n which is the space of homogeneous chaos of degree n and $\widehat{L^2(R^n)}$ into account.

It is not difficult (although only additional topological considerations are necessary) to have modifications of these facts to the case where we restrict the variables to an ovaloid (smooth, convex and closed surface) in R^n. This requirement is reasonable when a random field is indexed by an ovaloid.

Let m be taken specially to be $(n+1)/2$ in the above triple involving symmetric Sobolev spaces. Then, we have

$$\widehat{K^{(n+1)/2}(R^n)} \subset \widehat{L^2(R^n)} \subset \widehat{K^{-(n+1)/2}(R^n)},$$

and we form

$$H_n^{(n)} \subset H_n \subset H_n^{(-n)},$$

each space of this triple is isomorphic to the corresponding symmetric Sobolev space. More precisely, we recall the isomorphism I_n defined by (2.6.5), under which the H_n is isomorphic to $\widehat{L^2(R^n)}$, where the corresponding norms are different only by the constant $\sqrt{n!}$. Take the subspace $\widehat{K^{(n+1)/2}(R^n)}$ of $\widehat{L^2(R^n)}$, with a note that the former is equipped with a stronger topology. It is also noted that the isomorphism I can be restricted to any subspace, so that the restriction is denoted by the same symbol I. Let Q_n be the (continuous) injection such that

$$Q_n : \widehat{K^{(n+1)/2}(R^n)} \to \widehat{L^2(R^n)}.$$

Define

$$H^{(n)} = I_n^{-1} Q_n \widehat{H^{(n+1)/2}(R^n)}.$$

The $H^{(n)}$ is topologized so as to be isomorphic to the space $\widehat{K^{(n+1)/2}(R^n)}$. The norms of these two spaces are different only by the constant $\sqrt{n!}$.

Based on the basic scalar product in H_n, the dual space of $H^{(n)}$ can be defined in the usual manner, and it is denoted by $H^{(-n)}$.

The norms in those spaces are denoted by $\| \cdot \|_n, \| \cdot \|$, and $\| \cdot \|_{-n}$, respectively. Thus, we have a Hilbert space $H_n^{(-n)}$ of *generalized white noise functionals* of degree n.

There remains a freedom on how to sum up the spaces $H_n^{(-n)}, n \geq 0$, with suitable weights. Choose an increasing sequence $c_n > 0$, and form a Hilbert space $(L^2)^+$ by the direct sum:

$$(L^2)^+ = \bigoplus c_n H_n^{(n)}, \tag{2.6.6}$$

where $\varphi \in (L^2)^+$ is of the form

$$\varphi = \sum_n \varphi_n, \quad \varphi_n \in H_n,$$

and where the $(L^2)^+$-norm $\| \cdot \|$ is defined by

$$\|\varphi\|_+^2 = \sum_n c_n^2 \|\varphi_n\|_m^2.$$

The direct sum given above forms a Hilbert space and its dual space $(L^2)^-$ is expressed in the form

$$(L^2)^- = \bigoplus c_n^{-1} H_n^{(-n)}. \tag{2.6.7}$$

The definitions of $(L^2)^+$ and $(L^2)^-$ given above have rather formal significance, but they can show our idea behind the formality. The correct meaning is as follows.

The weighted algebraic sum $\sum c_n H_n^{(n)}$ is given without any difficulty. The topology is defined by the semi-norm $\|\cdot\|_+$. Let $(L^2)^+$ be the completion of the vector space spanned by those φ's that have finite semi norm. Eventually $\|\cdot\|_+$ becomes a norm. It is easy to see that $(L^2)^+$ is a separable Hilbert space with the Hilbertian norm $\|\cdot\|_+$.

Based on the (L^2)-norm, we can introduce a dual space $(L^2)^-$ of the Hilbert space $(L^2)^+$. Thus, each space $H_n^{(-n)}$ may be viewed as a subspace of $(L^2)^-$ and a member ψ of $(L^2)^-$ may be expressed as a sum of the form $\sum \psi_n, \psi_n \in H_n^{(-n)}$. We claim that $(L^2)^-$ is a Hilbert space with the norm $\|\cdot\|_-$ defined by $\|\psi\|_-^2 = \sum c_n^{-2} \|\psi_n\|_{-n}^2$.

Note. The choice of the sequence c_n depends on the problems to be discussed.

Naturally, we are given a triple

$$(L^2)^+ \subset (L^2) \subset (L^2)^-.$$

The space $(L^2)^+$ consists of test functionals, while $(L^2)^-$ is the space of *generalized white noise functionals*.

The S-transform can generalize the homeomorphism 2.5.5 to establish the space F^-, which is to be the extension of F in such a way that

$$(L^2)^- \sim F^{-1}. \tag{2.6.8}$$

Examples of $(L^2)^-$-functionals are given in the following.

Example 2.4 White noise $\dot{B}(t)$, and the renormalized square $: \dot{B}(t)^2 := \dot{B}(t)^2 - \frac{1}{dt}$.

White noise $\dot{B}(t)$ is a basic example from which we started to define generalized white noise functionals. Direct computation of $S(: \dot{B}(t)^2 :)$ is equal to $\xi(t)^2$. This result suggests one to consider the Wick products of $\dot{B}(t)$'s as will be discussed below.

Example 2.5 The Wick powers of $\dot{B}(t)$. We choose $:\dot{B}(t)^n:$. It belongs to $H_n^{(-n)}$. The associated S-transform is $\xi(t)^n$ and the kernel function is $\delta_t^{n\otimes}$. As for the definitions of the Wick product, one can refer to the book by Hida-Kuo-Potthof-Streit[64], but we briefly review some formulas.

Here are the formulas for the Wick product:

$$
\begin{aligned}
:1: &= 1 \\
:\dot{B}(u): &= \dot{B}(u) \\
:\dot{B}(u_1)\cdots\dot{B}(u_n): &= :\dot{B}(u_1)\cdots\dot{B}(u_{n-1}):\dot{B}(u_n) \\
&\quad -\sum_{i=1}^{n-1}\delta(u_n-u_i):\dot{B}(u_1)\cdots\dot{B}(u_{i-1})\dot{B}(u_{i+1})\cdots\dot{B}(u_n):.
\end{aligned}
$$

With these notations a member in H_n may be written as

$$\int\cdots\int F(u_1,\cdots,u_n):\dot{B}(u_1)\cdots\dot{B}(u_n):du_1\cdots du_n,$$

where F is in $\widehat{L^2(R^n)}$.

Note that the Wick products of $\dot{B}(t)$'s become simple products of $\xi(t)$'s by applying the S-transform. For instance

$$S(:\dot{B}(u_1)\cdots\dot{B}(u_n):)(\xi)=\prod\xi(u_i),\quad n=1,2,\cdots.$$

Incidentally, we can show that

$$\int_0^1 f(u):\dot{B}(u)^2:du,$$

f being smooth, is a member of $H_2^{(-2)}$. Its S-transform is

$$\int_0^1 f(u)\xi(u)^2du.$$

Example 2.6 $\delta_0(B(t))$ (the Donsker's delta function) is a member of $(L^2)^-$. Its S-transform is expressed simply in the form:

$$S(\delta_a(B(t)))(\xi)=\frac{1}{\sqrt{2\pi t}}\exp[-\frac{1}{2t}(a-\int_0^t\xi(u)du)^2].$$

This equation comes from the well-known formula of a delta function, namely

$$\delta_0(u) = \frac{1}{2\pi} \int e^{itu} dt.$$

Example 2.7 $\varphi_c(\dot{B}) = N \exp[c \int \dot{B}(t)^2 dt]$, $c \neq 1/2$, N being a normalizing constant. It is a generalized white noise functional that plays a role of the Gauss kernel in $(L^2)^-$. The S-transform of φ_c is given by

$$(S\varphi_c)(\xi) = \exp\left[\frac{c}{1-2c}\|\xi\|^2\right].$$

In reality the (multiplicative) renormalizing constant N is determined so that the S-transform is given by the above formula. cf. Kuo[93] Chapter 7.

Note. For $c > 0$, $\varphi_c(x)d\mu(x)$ defines a Gaussian measure, which is proved by the Bochner-Minlos theorem.

To close method A of constructing generalized white noise functionals, we give a remarkable note. Namely, the space $(L^2)^-$ enjoys a Hilbert space structure.

B. An analogue of the Schwartz space.

To fix the idea, we continue to take R to be the parameter space.

Take the operator A (see Section 2.3, Example 2.1):

$$A = -\frac{d^2}{du^2} + u^2 + 1$$

acting on $L^2(R)$. Then, apply the second quantization technique to introduce the operator $\Gamma(A)$ acting on the space $\bigoplus \widehat{L^2(R^n)}$ such that

$$\Gamma(A)\varphi = \sum_{n=0}^{\infty} I_n(A^{\otimes n} f_n)$$

where $I_n = \pi_n^{-1}$ (see (2.6.3)) and where $I_n(f_n)$ is the multiple Wiener integral of order n, the kernel f_n being in $\widehat{L^2(R^n)}$.

By using the isomorphism $\pi = \{\pi_n\}$ defined in **A** :

$$\pi_n : H_n \mapsto \widehat{L^2(R^n)},$$

we can easily define the operator $\tilde{\Gamma}(A) = \pi^{-1}\Gamma(A)\pi$. It is proved, as in **A**, that for $\varphi \in H_n$

$$\|\varphi\| = \sqrt{n!}\|\pi\varphi\|_{L^2(R^n)}.$$

If no confusion occurs, $\tilde{\Gamma}(A)$ is also denoted by $\Gamma(A)$. Note that it acts on (L^2).

Set $(S)_n = \mathcal{D}(\Gamma(A)^n)$ and set $\|\varphi\|_n = \|\Gamma(A)^n\varphi\|$. Define the space (S) by

$$(S) = \bigcap_n (S)_n.$$

The projective limit topology is introduced in (S). This space (S) is taken to be the space of *test functionals*.

Example 2.8 A functional $\exp[a\langle x, \xi\rangle]$, $a \in \mathbf{C}$, is a test functional.

Proposition 2.5 *The space (S) is a complete topological vector space and it is an algebra.*

Proof. See Hida-Kuo-Potthof-Streit [64].

■

The dual space of $(S)_n$ is denoted by $(S)_{-n}$ and the family $\{(S)_{-n}, n \geq 0\}$ forms a consistent system; namely $(S)_{-n} \subset (S)_{-(n+1)}$with Hilbert-Schmidt type injection, and the injections $(S)_{-n} \mapsto (S)_{-(n+1)}, n \geq 0$, are consistent. We can therefore define the projective limit:

$$(S)^* = \text{proj} \cdot \lim_{n\to\infty} (S)_{-n}.$$

Definition 2.7 The projective limit space $(S)^*$ is the space of *generalized white noise functionals*. It is often called the space of *white noise distributions* (see Kuo[93]).

Note. The functional $\exp[\langle x, \xi\rangle]$ is in (S). This fact implies that the S-transform can be extended to an operator acting on $(S)^*$.

There is a characterization of $(S)^*$-functional in terms of S-transform by J. Potthoff and L. Streit[126]. This is one of the significant advantages of using the space $(S)^*$.

First the notion of a *U-functional* is reminded. To fix the idea, the basic nuclear space is taken to be the Schwartz space S. Note that (S) is topologized by the sequence of Hilbertian norms $\|\xi\|_n = \|A^n\xi\|$.

Definition 2.8 Let $U(\xi)$ be a complex-valued functional on S. U is called a U-functional if the following two conditions are satisfied.

1) For all $\xi, \eta \in S$, the mapping $\lambda \mapsto U(\eta + \lambda\xi)$, $\lambda \in R$ has an entire analytic extension, denoted by $U(\eta + \lambda\xi), \lambda \in \mathbf{C}$,
2) there exist $p \in N$ and $c_1, c_2 > 0$ such that for all $z \in \mathbf{C}, \xi \in S$, the inequality

$$|U(z\xi)| \le c_1 \exp[c_2|z|^2|\xi|_p^2]$$

holds.

We now state the theorem due to Potthoff and Streit[126].

Theorem 2.4 *(Potthoff-Streit[126]) If $\varphi \in (S)^*$, then $S\varphi$ is a U-functional. Conversely, if U is a U-functional, then there exists a unique φ in $(S)^*$ such that $S\varphi = U$.*

This criterion for φ to be in $(S)^*$ is quite useful in many occasions. For instance, suppose we have solved a variational equation in terms of S-transform to have a solution functional $U(\xi)$. Then, it is necessary to check if the solution is really a S-transform of some (generalized) white noise functional. The above theorem is quite helpful in such cases.

Proposition 2.6 *$\ddot{B}(t)$ or $\ddot{x}(t)$, $x \in E^*(\mu)$, is in $(S)^*$.*

Proof. We see that the S-transform of $\ddot{B}(t)$ is well defined and its characteristic functional is of the form

$$\begin{aligned}(S\ddot{B})(\xi) &= C(\xi) \int_{E^*} \exp[\langle x, \xi\rangle]\dot{x}(t)d\mu(x) \\ &= \frac{d}{dt}C(\xi) \int_{E^*} \exp[\langle x, \xi\rangle]x(t)d\mu(x)\end{aligned}$$

$$= \xi'(t).$$

The evaluation

$$|\xi'(t)| = \left|\int_{-\infty}^{t} \xi''(u)du\right|$$
$$\leq \int_R (1+u^2)^{-1/2}(1+u^2)^{1/2}|\xi''(u)|du$$
$$\leq \sqrt{\int (1+u^2)^{-1}du \int (1+u^2)|\xi''(u)|^2 du},$$

proves that

$$c\|\xi\|_2 \leq c\exp[\|\xi\|_2].$$

Hence, we have the assertion by using Theorem 2.4.

■

To add the above proposition, we should like to state from a different viewpoint.

Given a functional $U(\xi) = \xi(t)$, where t is fixed. Then, we can prove thast it is a U-functional. In addition, it is the time derivative of ξ, so that the associated white noise functional is the derivative of functional associated to $\dot{B}(t)$. Namely, we have to have $\ddot{B}$ corresponding to the given functional $\xi'(t)$.

We have an observation.

First we note that $\ddot{B}(t), t \in R$, is a generalized stochastic process with *independent values at every point* in the sense of Gel'fand-Vilenkin[30]. Its characteristic functional is given as follows:

$$E\{\exp[i\langle\ddot{B}, \xi\rangle]\} = E\{\exp[-i\langle\dot{B}, \dot{\xi}\rangle]\},$$

which is equal to

$$\exp[-\frac{1}{2}\|\dot{\xi}\|^2].$$

The probability distribution μ_1 of $\ddot{B}(t)$ is introduced in the space E^* similar to the case of $\dot{B}(t)$, although their supports are different.

Similar to $\dot{B}(t)$, $\ddot{B}(t)$ has two faces; one is that a single $\ddot{B}(t)$ has its own identity as a member of $(S)^*$, and the other is that the $\ddot{B}(t), t \in R$ is a generalized stochastic process. Indeed, it is stationary in time.

On the other hand, we can easily see that $\ddot{B}(t)$ is not in $H_1^{(-1)}$, so that it is not a generalized functional in $(L^2)^-$. In fact, if it were in $(L^2)^-$, then it should be a linear functional of $\dot{x}$. Of course, it is not the case. Incidentally, the kernel function of the integral representation (that is the kernel of the U-functional) is $-\delta_t(u)$, so that it is not in the Sobolev space $H^{-1}(R^1)$. Because the Fourier transform of the kernel is $-i\lambda e^{it\lambda}$, which fails to be square integrability of the multiplication by $(1+\lambda^2)^{1/2}$.

Example 2.9 It can be easily seen that

$$: \ddot{B}(t)^n : \in (S)^*$$

and

$$B^{(k)}(t) \notin (L^2)^-, \; k \geq 2$$

$$B^{(k)}(t) \in (S)^*, \qquad \text{for every } k,$$

where $B^{(k)}(t) = \frac{d^k}{dt^k} B(t)$ in the distribution sense.

$$(L^2) \subset (L^2)^{-1}$$
$$(L^2) \subset (S)^* \subset L^2(\mu)_{-\infty}$$

Diagram 2.2.

Some time later, W.G. Cochran, H.-H. Kuo and A. Sengupta introduced a new space $L^2(\mu)_{-\infty}$ of generalized functionals much wider than $(S)^*$. The corresponding characterization is given in a similar manner, but the inequality for U-functional is naturally generalized. We have to skip the details and refer to their paper[20] for details. We note however that the new class is quite significant in our analysis; for example it suggests some connection with the Poisson noise.

Also, we would like to recall that Yu.G. Kondratiev and L. Streit have discussed another direction of generalization.

Before closing this chapter, we would like to note additional interesting properties of the space (S). They will be referred in Chapter 8.

Proposition 2.7 *Each member of (S) has a continuous version.*

A positive generalized functional is defined by

Definition 2.9 A generalized white noise functional $\varphi(x)$ is said to be *positive* if

$$\langle \varphi, \psi \rangle \geq 0,$$

for any positive test functional ψ.

To prove these assertions, several steps are necessary, so we refer to the literature[90] by I. Kubo and Y. Yokoi. Positivity will be used in Chapter 10, Section 3.

The collection of positive generalized white noise functionals in $(S)^*$ will be denoted by $(S)^*_+$.

Concerning positive generalized functionals, further properties can be seen in Y. Yokoi[169].

2.7 Creation and annihilation operators

Once again we emphasize the basic idea of our analysis that we introduce the space $(L^2)^-$ of generalized white noise functionals since $\dot{B}(t)$ (or $x(t)$) is a variable of white noise functionals.

Now it is quite reasonable to define partial derivatives in the variables $\dot{B}(t)$'s.

First, we wish to introduce a partial differential operators expressed symbolically in the form

$$\partial_t = \frac{\partial}{\partial \dot{B}(t)}.$$

Since the variable $\dot{B}(t)$ is a generalized functional in $H_1^{(-1)}$, the partial differential operator is defined in a generalized sense as is seen in Definition 2.10 below.

Note. The notation ∂_t is due to Kubo-Takenaka[89]; the same for the adjoint operator ∂_t^*.

Incidentally, we claim that the operator ∂_t should be extended to an *annihilation operator* acting on the space of generalized white noise functionals. Rigorous definition is given as follows.

Definition 2.10 Let φ be a generalized white noise functional and let $U(\xi)$ be its S-transform. Then the *annihilation operator* is defined by

$$\partial_t \varphi = S^{-1} \frac{\delta}{\delta \xi(t)} U(\xi), \tag{2.7.1}$$

if $\frac{\delta}{\delta \xi(t)} U(\xi)$ exists and is a U-functional. The notation $\frac{\delta}{\delta \xi(t)}$ stands for the Fréchet derivative.

Intuitively speaking, the Fréchet derivative may be viewed as a *continuous* analogue of a differential of a function $u = U(x_1, x_2, \cdots, x_n)$ on R^n, such that

$$du = \sum_1^n \frac{\partial u}{\partial x_i} dx_i,$$

where each coordinate (or direction) contributes eaqualy in $i, 1 \le i \le n$. Similarly each $\dot{B}(t)$, where $t \in R$, contributes equally in t in our analysis. We shall be back to this understanding later in Chapter 5.

Remark 2.5 *In order to define a partial derivative, there is a method of using Gâteaux derivative. Namely, taking a complete orthonormal system $\{\xi_n\}$ of $L^2(R)$, with $\xi_n \in E$ for every n, we introduced the operator ∂_n in Section 2.5. It is also an annihilation operator, but is not quite convenient for our purpose, since the parameter t disappears. This is far from continuously many dimensional (see Lévy's construction of a Brownian motion explained in Section 2.1, the item (4) of construction of a Brownian motion).*

Proposition 2.8 *The domain of annihilation operator ∂_t includes $(L^2)^+$ and ∂_t is a continuous surjection such that*

$$\partial_t : H_{n+1}^{(n+1)} \mapsto H_n^{(n)}. \tag{2.7.2}$$

Proof. The S-transform of $\varphi \in H_n^{(n)}$ is of the form

$$\int \cdots \int F(u_1, \cdots, u_n)\xi(u_1)\cdots\xi(u_n)du^n;$$

where $F \in \widehat{K^{\frac{n+1}{2}}}(R^n)$. Apply $\frac{\delta}{\delta\xi(t)}$ to obtain a kernel $F(u_1, \cdots, u_{n-1}, t)$. By the trace theorem for the Sobolev space, we prove that $F(u_1, \cdots, u_{n-1}, t)$ is in $\widehat{K^{\frac{n}{2}}}(R^{n-1})$. Kernel functions of this form runs through whole space $K^{\frac{n}{2}}(R^{n-1})$ for every t. This proves the assertion. ∎

Corollary 2.1 *The creation operators ∂_t^* acts in such a way that*

$$\partial_t^* : H_n^{(-n)} \mapsto H_{n+1}^{(-n-1)}.$$

The mapping ∂_t^* is a continuous injection.

Proof. Computation of the Sobolev norms of the kernel functions of order $-n$ and $-n-1$ prove the assertion. ∎

The operator ∂_t has a domain which is rich enough, including the space (S) of test functionals, so that we can define its adjoint operator ∂_t^* by

$$\langle \partial_t^* \varphi, \psi \rangle = \langle \varphi, \partial_t \psi \rangle,$$

for $\varphi \in (S)^*$ and $\psi \in (S)$.

These operators extend to those with the domain $(L^2)^{-1}$ for (L^2).

Corollary 2.2 *The classical stochastic integral $\int \varphi(t,x)dB(t)$ is expressed in the form $\int \partial^* \varphi(t,x)dt$.*

The operator ∂_t^* is used to define the so-called *Hitsuda-Skhorohod integral*, where the integrand are not necessarily non-anticipating. Hence, the integral has various applications.

The integral is of the form

$$\int \partial_t^* \varphi(x)dt$$

where φ is not necessarily to be non-anticipating.

Example 2.10 If $\varphi(x) = e^{\langle x,\xi\rangle - \frac{1}{2}\|\eta\|^2}$, then

$$(S\varphi)(\xi) = e^{\langle \eta,\xi\rangle}.$$

Its Fréchet derivative is

$$\frac{\delta}{\delta\xi(t)} e^{\langle \eta,\xi\rangle} = \eta(t)e^{\langle \eta,\xi\rangle},$$

and hence,

$$\partial_t \varphi(x) = \eta(t)\varphi(x).$$

Theorem 2.5 *The following assertions hold.*

i) The domain $\mathcal{D}(\partial_t)$ includes (S).
ii) The ∂_t is an annihilation operator, in particular it is a surjection of (S) onto itself.
iii) For $\varphi \in H_n^{(n)}$ such that the kernel $F(u_1, \cdots, u_{n-1}, t)$ is associated with $(S\partial_t\varphi)(\xi)$.

Proof. Assertions are easily proved. For example, if φ is in $H_n^{(n)}$, then the kernel function φ is of the form $F(u_1, \cdots, u_n)$ which is symmetric. Hence $nF(t, u_1, \cdots, u_{n-1})$ is associated with $\partial_t\varphi$. The kernel theorem for Sobolev space claims that $nF(u_1, \cdots, u_{n-1}, t)$ is again in the Sobolev space $K^{\frac{n}{2}}\widehat{(R^{n-1})}$. Now follow the assertions.

■

Theorem 2.6 *The following assertions are true.*

i) The domain $\mathcal{D}(\partial_t^*)$ is dense in $(S)^*$ and $(L^2)^-$.
ii) The ∂_t^* is a creation operator such that

$$(S\partial_t^*\varphi)(\xi) = \xi(t)(S\varphi)(\xi). \tag{2.7.3}$$

Note. This theorem is a counter part of Theorem 2.5.

2.8 Examples

These are examples of application. Applications to Physics will be discussed in Chapter 10. In this section, we only deal with typical applications to stochastic differential equations.

Example 2.11 Consider a stochastic differential equation which is expressed in the traditional form;

$$dX(t) = (aX(t) + a')dt + (bX(t) + b')dB(t). \tag{2.8.1}$$

Both coefficients of the terms dt and $B(t)$ are linear. We may assume $E[X(t)] = 0$, so that $a' = 0$. The essential data at $t = t_0$ is taken to be a constant, say $X(t_0) = C$.

Now the second term on the right-hand side is the multiplication by $dB(t)$ with $dt > 0$, which is independent of the $X(s), s \le t$. Thus, it can be replaced by $\partial_t^*(bX(t) + b')$ without taking any notice of non-anticipating property. Thus, the $X(t)$ is assumed to be a function of $\dot{B}(s), s \le t$, and the given equation can be expressed in the form

$$\frac{d}{dt}X(t) = aX(t) + \partial_t^*(bX(t) + b'). \tag{2.8.2}$$

Applying the S-transform and using the notation

$$\begin{aligned} \frac{d}{dt}U(t,\xi) &= aU(t,\xi) + \partial_t^*(bU(t,\xi) + b') \\ &= (a + b\xi(t))U(t,\xi) + b'\xi(t), \\ U(t_0,\xi) &= C. \end{aligned} \tag{2.8.3}$$

For a fixed ξ, this is an elementary linear differential equation, so that we immediately obtain the solution $U(t,\xi)$.

$$U(t,\xi) = e^{\int_{t_0}^t (a+b\xi(s)ds} \left(C + \int_{-\infty}^t b'\xi(s) e^{-\int_{t_0}^s (a+b\xi(u)du} \right) ds. \tag{2.8.4}$$

This satisfies the Potthoff-Streit condition so that $U(t,\xi)$ is an U-functional. Hence

$$X(t) = S^{-1}U(t,\xi)$$

is the solution of the given stochastic differential equation.

To fix the idea, we let the solution be a stationary stochastic process. Letting t_0 tend to $-\infty$, and $C = 0$.

Hence we have

$$U(t,\xi) = e^{at}e^{b\int_{-\infty}^{t}\xi(s)ds}\int_{-\infty}^{t} b'\xi(s)e^{-as}e^{-b\int_{-\infty}^{s}\xi(u)du}ds$$

$$= b'\int_{-\infty}^{t} e^{a(t-s)}\xi(s)e^{b\int_{s}^{t}\xi(u)du}ds. \tag{2.8.5}$$

Let $U(t,\xi) = \sum_{1}^{\infty} U_n(t,\xi)$ be the power series expansion, where $U_n(t,\xi)$ is a polynomial in ξ of degree n.

Then, we have

$$U_n(t,\xi) = b'\int_{-\infty}^{t} e^{a(t-s)}\xi(s)\frac{b^{n-1}}{(n-1)!}\left(\int_{s}^{t}\xi(u)du\right)^{n-1} ds. \tag{2.8.6}$$

Let $F_n(u_1,\cdots,u_n)$ be the kernel function of $U_n(0,\xi)$, which has to be symmetric, such that

$$U_n(t,\xi) = \int^{0}\cdots\int^{0} F_n(u_1,\cdots,u_n)du_1\cdots du_n.$$

Changing the order of integration (2.8.6) and symmetrize, we have

$$F_n(u_1,\cdots,u_n) = \frac{1}{n!}b^{n-1}b'\exp\left[-a\min_{1\le j\le n} u_j\chi_{(-\infty,0]^n}(u_1,\cdots,u_n)\right].$$

Summing up, the solution of the equation (2.8.3) is given by $U_n(t,\xi), n = 1,2,\cdots$ with the kernel $F_n(u_1-t,\cdots,u_n-t)$. The solution $X(t)$ of (2.8.2) can be expressed as

$$X(t) = \sum_{n=1}^{\infty} X_n(t), \qquad X_n(t) = S^{-1}U_n(t,\xi).$$

This is in agreement with the process discussed in Hida[40] Section 4.6, Example 2.

Example 2.12 We consider a modified Langevin equation of the form

$$\frac{d}{dt}X(t) = -\lambda X(t) + a :\dot{B}(t)^2: \tag{2.8.7}$$

where the fluctuation is a normalized square of white noise expressed by Wick product.

Assume that $\lambda > 0$ and a is an arbitrary non-zero constant. For a moment, we put the initial condition

$$X(t_0) = C_0.$$

To solve the equation, computation should be done within the space $(L^2)^-$ of generalized white noise functionals. Set

$$SX(t) = U(t, \xi).$$

Then, equation (2.8.7) turns into

$$\frac{d}{dt}U(t,\xi) = -\lambda U(t,\xi) + a\xi(t)^2$$
$$U(t_0, \xi) = C_0.$$

It is easy to solve the equation to obtain

$$U(t,\xi) = e^{-\lambda(t-t_0)}\left(C_0 + a\int_{t_0}^{t} e^{\lambda s \xi(u)^2 du}\right). \tag{2.8.8}$$

As in the last example, we wish to have a stationary solution by letting $t_0 \to -\infty$ and $C_0 \to 0$.

Finally, we have

$$U(t,\xi) = a\int_{-\infty}^{t} e^{-\lambda(t-u)\xi(u)^2} du.$$

Hence, the solution $X(t)$ of equation (2.8.7) is expressed in the form

$$X(t) = a\int_{-\infty}^{t} : e^{-\lambda(t-u)\dot{B}(u)^2} : du.$$

The $X(t)$ runs through the space $(L^2)^{-1}$.

Example 2.13 Again we discuss within the space $(L^2)^-$ or $(S)^*$. Consider the equation

$$\partial_t \varphi(x) = a\partial_t^* \varphi(x), \tag{2.8.9}$$

where a is a non-zero constant.

For $U(\xi) = (S\varphi)(\xi)$, we have

$$\frac{\delta}{\delta\xi(t)}U(\xi) = a\xi(t)U(\xi). \tag{2.8.10}$$

To solve the equation, we expand $U(\xi) = \sum_1^\infty U_n(\xi)$, where $U_n(\xi)$ is homogeneous in ξ. The U_n's are arranged so that the degree is increasing.

By assumption, $U_0(\xi) = 1$. As for $U_1(\xi)$, we are given

$$\frac{\delta}{\delta\xi(t)} U_1(\xi) = a\xi(t)U_0(\xi).$$

The choice of U_0 on the right-hand side comes from the counting degree. The U_1 has to be quadratic. In fact, we have

$$U_1(\xi) = \frac{a}{2}\int \xi(u)^2 du.$$

In general, we must have

$$\frac{\delta}{\delta\xi(t)} U_n(\xi) = a\xi(t)U_{n-1}(\xi), n = 1, 2, \ldots.$$

Hence

$$U(\xi) = \frac{1}{2}e^{a\int \xi(u)^2 du}.$$

This is the U-transform of the Gauss kernel

$$\varphi_c(x) = Ne^{c\int x(u)^2 du},$$

where $c = \frac{a}{2a+1}$ and N is the renormalizing constant. The $\varphi_c(x) = \varphi_c(\dot{B})$ has been discussed in Example 2.7.

The φ_c will be discussed later in connection with the Lévy Laplacian.

2.9 Addenda

A.1. The Gauss transform, the S-transform and applications

In the Princeton Lecture Notes[36], one of the authors has discussed the Fourier-Wiener transform, following Cameron-Martin, in connection with the Fourier-Hermite polynomials. Also a definition of Gauss transform is given there. Its infinite dimensional analogue can be considered and even suggests us basic concepts in white noise analysis.

We now define the *Gauss transform* on (L^2) of a white noise functional φ (maybe generalized functional) by the formula (an analogue of the classical

Gauss transform, (see e.g. Higher transcendental functions. Vol. 2, McGraw-Hill, 1953). We modify the classical formula as folows so as to be fitting for the present calculus.

$$G\varphi(\xi) = \int \varphi(x + \xi) d\mu(x).$$

Remind that

$$\frac{d\mu(x+\xi)}{d\mu(x)} = \exp[\langle x, \xi\rangle - \frac{1}{2}\|\xi\|^2],$$

we obtain

$$G\varphi(\xi) = e^{-\frac{1}{2}\int e^{\langle x,\xi\rangle \varphi(x) d\mu(x)}} = (S\varphi)(\xi).$$

Again, by modifying the Gauss transform to $\tilde{G}$:

$$\tilde{G}\varphi(\xi) = \int \varphi(x + i\xi) d\mu(x),$$

we are given the formula for the Wick product of monomial in $\dot{B}(t)$'s.

Note. Y. Yokoi [Hiroshima J. 1995] discussed a transform G' of white noise functionals φ:

$$G'\varphi(y) = \int_{E^*} \varphi(x + \frac{1}{\sqrt{2}} y) d\mu(x),$$

which is called Gauss transform. Its inverse is of the form

$$(G')^{-1} f(x) = \int_{E^*} f(\sqrt{2}(x + iy)) d\mu(y).$$

Similar results to these and beyond are obtained.

Remark 2.6 *It is important to recongnize that the S-transform is quite different from the classical Bargmann-Segal type transformation as we can see in many places in what follows.*

The S-transforms of Fourier-Hermite polynomials.

Recall the formulae of the (ordinary) Hermite polynomials $H_n(x), n \geq 0$:

$$H_n(x) = (-1)^n e^{x^2} \frac{d^n}{dx^n} e^{-x^2}.$$

The generating function of the $H_n(x)$ is

$$\sum_{n=0}^{\infty} \frac{t^n}{n!} H_n(x) = \exp[-t^2 + 2tx].$$

We have

Proposition 2.9 *Let f be a unit vector in $L^2(R)$. The S-transform of $2^{-n/2} H_n(\frac{\langle x, f\rangle}{\sqrt{2}})$ is $\langle \xi, f\rangle^n$.*

Proof. We prove by computing the S-transform of generating function with variable $\langle x, f\rangle$:

$$S(\epsilon^{-t^2/2 + t\langle x, f\rangle})(\xi) = \exp[t\langle \xi, f\rangle].$$

∎

A.2. The Karhunen-Loève expansion

Suppose we are given a continuous, symmetric, positive definite function $\Gamma(s,t), (s,t) \in [0,1]^2$, such that

$$0 < \int_0^1 \Gamma(t,t)dt < \infty.$$

Then, there exists a decreasing sequence $\{\lambda_k \geq 0; k \geq 1\}$, together with a system of functions $\{e_k(t); k \geq 0\}$, with $e_k(t) \in L^2[0,1]$ for every k, such that the following conditions (1)-(4) hold :

(1) For any $i, k \geq 1$

$$\int_0^1 e_i(t) e_k(t) dt = \begin{cases} 0, & i \neq k, \\ n, & i = k. \end{cases} \tag{2.9.1}$$

(2) The system $\{\lambda_k, e_k; k \geq 1\}$ forms a complete set of solutions of the equation

$$\lambda e(t) = \int_0^1 \Gamma(t,s) e(s) ds; 0 < t < 1,$$

with $\int_0^1 e(t)^2 dt = 1$.

(3) The series (t,s)

$$\sum_{k=1}^{\infty} \lambda_k e_k(t) e_k(s), \quad 0 < s,t < 1,$$

converges in $L^2([0,1]^2)$ and the sum is equal to $\Gamma(s,t)$.

(4) For a system $\{X_k\}$ of independent random variable subject to the standard Gaussian distribution $N(0,1)$ we have the Karhunen-Loéve expansion :

$$X(t) = \sum_{k \geq 1} \sqrt{\lambda_k} X_k e_k(t),$$

which always converges almost surely to have a Gaussian process and the covariance function is in agreement with $\Gamma(s,t)$.

A.3. Reproducing kernel Hilbert space

Let E be any non-empty abstract space, and let $K(\xi,\eta), (\xi,\eta) \in E \times E$, be real positive definite; namely, for any n and any $\xi_1, \xi_2, \cdots \xi_n \in E$, and any complex numbers $\alpha_1, \alpha_2, \cdots, \alpha_n$, the inequality

$$\sum_{j,k=1}^{n} K(\xi_j, \xi_k)\alpha_j \bar{\alpha}_k \geq 0$$

holds.

The following properties i), ii) and iii) for a positive definite function $K(\xi,\eta)$ are known.

i) $K(\xi,\eta) \geq 0$

ii) $K(\xi,\eta) = K(\eta,\xi)$

iii)$|K(\xi,\eta)|^2 \leq K(\xi,\xi)K(\eta,\eta)$.

Theorem 2.7 *Let $K(\xi,\eta), (\xi,\eta) \in E \times E$, be real positive definite. Then, there exists a unique real Hilbert space* $\mathbf{F}$ *such that*

i) $(f(\cdot), K(\cdot,\eta)) = f(\eta)$

ii) $\mathbf{F}$ *is generated by* $K(\cdot,\eta), \eta \in E$.

Definition 2.11 The Hilbert space $\mathbf{F}$ satisfying the properties i) and ii), is called a *reproducing kernel Hilbert space* (RKHS) with kernel $K(\xi,\eta)$.

Proof of the theorem. Define a collection $\mathbf{F}_1$ defined by

$$\mathbf{F}_1 = \left\{ \sum_{k=1}^{n} a_k K(\xi, \eta_k); a_k \in R, \xi, \eta_k \in E \right\}.$$

Obviously $\mathbf{F}_1$ is a vector space. We define a semi-norm $\| \ \|_1$ by

$$\| \sum_{k=1}^{n} a_k K(\cdot, \eta_k) \|_1^2 = \sum_{k=1}^{n} a_j a_k K(\eta_j, \eta_k).$$

Obviously $\| \ \|_1$ is a Hilbertian semi-norm. Define a subspace N of $\mathbf{F}_1$ by,

$$N = \left\{ f = \sum_{k=1}^{n} a_k K(\cdot, \eta_k), \|f\|_1 = 0 \right\},$$

which is proved to be a vector space. The mod N defines equivalent classes i.e. there exists a factor space $\mathbf{F}/N$, which is denoted by $\mathbf{F}$. A Hilbertian norm $\| \ \|$ is defined on $\mathbf{F}$.

The class, a representative of which is $K(\cdot, \eta)$, is denoted by the same symbol $K(\cdot, \eta)$. The same for linear combinations of $K(\cdot, \xi)$'s. With this notation, we can show the reproducing property

$$(f(\cdot), K(\cdot, \xi) = f(\xi),$$

where $(\cdot, \cdot)$ denote the inner product defined by $\| \ \|$.

Now, if necessary, let $\mathbf{F}$ be complete. (The completion is done by the usual method.) The reproducing property still holds.

Finally, we show the uniqueness. Suppose there were another reproducing kernel $K'(\xi, \eta)$, defined from $K(\xi, \eta)$, in $\mathbf{F}$. Then

$$\begin{aligned} 0 &< \|K(\xi, \eta_0) - K'(\xi, \eta_0)\|^2 \\ &= (K(\xi, \eta_0) - K'(\xi, \eta_0), K(\xi, \eta_0) - K'(\xi, \eta_0)) \\ &= \{K(\eta_o, \eta_0) - K'(\eta_0, \eta_0)\} - \{K(\eta_0, \eta_0) - K'(\eta_0, \eta_0)\} \\ &= 0, \end{aligned}$$

where the "reproducing kernel property" is used.

This is a contradiction.

■

Note. There exists a reproducing kernel in a Hilbert space $\mathbf{F}$ if and only if

$$|f(\eta_0)| \le C_{\eta_0} \|f\|$$

with a constant C_{η_0}, independent of f, i.e. $f(\eta_0)$ is a continuous linear functional of f for any $\eta_0 \in E$.

Remark 2.7 *If the kernel $K(\xi, \eta)$ is complex valued, we can also form a complex Hilbert space* $\mathbf{F}$ *in a similar manner, but some modification of the inner product is necessary.*

Chapter 3

Elemental random variables and Gaussian processes

The term "elemental" in the title of this chapter may sound unfamiliar, but it means "basic" and "atomic". It is often called elementary. Recall the reductionism explained in Chapter 1. As far as random system is concerned, we want to find mutually independent and atomic elements, in terms of which the random system in question is composed of. Such an elemental system, if it is stationary in time and is an independent system, is said to be *idealized,* in fact, the system becomes a stationary generalized stochastic process (or a generalized random field) in the case of a process (or a random field) with continuous parameter.

3.1 Elemental noises

There are two typical systems of idealized elemental random variables; one is Gaussian noise (that is white noise) and the other is Poisson noise which is the time derivative of a Poisson process. They serve to prepare the background of our analysis, and we shall, therefore, start with a brief review of those noises.

The main aim in this chapter is to establish a theory of linear functionals of those noises. The theory proceeds with different flavor from that of Chapter 2 depending on the respective cases. The definitions of those noises are given in the previous chapter, however, it would be useful to remind the notion of a random measure which could give an intuitive interpretations when we discuss linear functionals or stochastic integrals based on them. Although the roles of a random measure are familiar to us in defining linear functionals, we should mention shortly so as to make these notes self-contained, including even more general cases without restricting either to Gaussian or Poisson noise.

To fix the idea, the time parameter space is taken to be R^d. Let $(R^d, \mathcal{B}, v)$ be a locally finite measure space, where $\mathcal{B}$ is the sigma-field of Borel subsets of R^d.

I. The first method of stochastic integral.

To fix the idea, the time parameter space is taken to be R^d. Let $(R^d, \mathcal{B}, v)$ be a locally finite measure space, where $\mathcal{B}$ is the sigma-field of Borel subsets of R^d.

Definition 3.1 A system of real valued random variables

$$\mathbf{Y} = \{Y(B) = Y(B, \omega), B \in \mathcal{B}\}$$

on a probability space $(\Omega, \mathbf{B}, P)$ is called a *random measure*, if $\mathbf{Y}$ satisfies

i) $Y(B_1)$ and $Y(B_2)$ are independent for any $B_1, B_2 \in \mathcal{B}$ such that $B_1 \bigcap B_2 = \emptyset$,

ii) each $Y(B)$ has finite moments up to second order and

$$E(Y(B)) = 0, \quad E(Y(B)^2) = v(B),$$

iii) finitely additive property holds, i.e. for a disjoint sum $B = \sum_{i=1}^n B_i, B \in \mathcal{B},\ v(B_i) < +\infty$, the inequality

$$\sum_{i=1}^n Y(B_i) = Y(B)$$

holds almost surely.

The completely additive property is easily implied from finitely additive property by noting that v is a measure and by using independent property for disjoint Borel subsets.

Remark 3.1 *A complex-valued random measure can also be defined in the same manner provides that the variance in ii) above is replaced by* $E(|Y(B)|^2) = v(B)$.

We are now ready to define a stochastic integral based on a given (real) random measure $\mathbf{Y}$ in a similar manner to the Lebesgue integral.

1) Take a complex valued step function

$$f(t) = \sum_{i=1}^{n} a_i \chi_{B_i}(t)$$

with $v(B_i) < \infty$ and $a_i \in \mathbf{C}$. Define a stochastic integral

$$\int f(t)Y(dt) = \sum_{i=1}^{n} a_i Y(B_i). \tag{3.1.1}$$

2) For f in $L^2(R^d, v)$, choose a sequence of step functions f_k that converges to f in the L^2 sense. Then, we have $\int f_n(t)Y(dt)$, which is denoted by X_n. That is

$$X_n = \int f_n(t)Y(dt). \tag{3.1.2}$$

We prepare the complex Hilbert space $L^2(\Omega, P)$, the norm of which is denoted by $\|\cdot\|$.

Proposition 3.1 *The X_k's expressed in (3.1.2) form a Cauchy sequence in the topology defined by $\|\cdot\|$, so that strong limit of the X_k exists in the Hilbert space $L^2(\Omega, P)$. The limit does not depend on the choice of the sequence f_n that approximates f, but depends only on f. Namely, the integral is well defined.*

Proof. We have

$$E(|X_j - X_k|^2) = \int |f_j(t) - f_k(t)|^2 dv(t).$$

Since the f_k's form a Cauchy sequence in $L^2(R^d, \mathcal{B}, v)$, so does the sequence X_k's in $L^2(\Omega, P)$. Hence, the sequence X_k tends to a certain member, denoted by X, in $L^2(\Omega, P)$. It is not hard to show that the limit X does not depend on the choice of the sequence f_j, but depends only on f, so that X is written as $X(f)$.

Following the notation introduced in 1), we may write the $X(f)$ as the integral based on the random measure Y. Thus, we have

$$X(f) = \int f(t)Y(dt).$$

This integral is also written as

$$X(f) = \int f(t)dY(t).$$

■

Definition 3.2 The above $X(f)$ is called a *stochastic integral* of f based on random measure Y.

A stochastic integral is continuous and linear in the integrand f.

Let $\mathbf{M}(Y)$ be the subspace of the Hilbert space $L^2(\Omega, P)$ spanned by $X(f), f \in L^2(R^d, v)$. Since it holds that $\|X(f)\|^2 = \int |f(t)|^2 d\nu(t)$, we can prove the isomorphism

$$\mathbf{M}(Y) \cong L^2(R^d, v). \tag{3.1.3}$$

Example 3.1 White noise integral is a stochastic integral based on white noise $\dot{B}(t)$, for the case $d = 1$.

Take the measure v to be the Lebesgue measure, and set $Y(dt) = \dot{B}(t)dt = dB(t)$. The integral defined above is often written in the form

$$\int f(t)dB(t), \tag{3.1.4}$$

or

$$\int f(t)\dot{B}(t)dt. \tag{3.1.5}$$

We shall later prefer the expression (3.1.5) to (3.1.4) for computational reason and others. The space $\mathbf{M}(B)$ is in agreement with H_1 defined in Chapter 2.

Example 3.2 Poisson noise integral, where $d = 1$.

Take a Poisson process with parameter space R^1. Then, replacing B with P in (3.1.4) and (3.1.5), we establish the stochastic integral based on a Poisson noise $\dot{P}(t)$. We are given a continuous, linear functional of Poisson noise. It is written as

$$\int f(t)dP(t), \qquad \int f(t)\dot{P}(t)dt,$$

and the space $\mathbf{M}(P)$ is defined.

Details will be discussed later.

II. The second method of stochastic integral.

First we restrict our attention to *Gaussian random measure*, that is $\{Y(B), B \in \mathcal{B}, v(B) < \infty\}$, B being a Borel subset of R^d, is a Gaussian system. This means that any finite linear combination of $Y(B)$'s is subject to a Gaussian distribution, in particular, we assume that the probability distribution of $Y(B)$ is $N(0, v(B))$, v being a Borel measure which is a positive tempered distribution. More details of a Gaussian system will be explained in the next section.

By method I, the collection $H = \{X(f), f \in L^2(R^d, v)\}$ forms a Gaussian system. Now we come to the next method.

For $\xi \in E$, E being a nuclear space, say the Schwartz space over R^d, we have

$$C_v(\xi) = E(\exp[i \int \xi(t) Y(dt)]) = \exp[-\frac{1}{2} \int \xi(t)^2 v(dt)],$$

which is a characteristic functional of a Gaussian measure, denoted by μ_v. It is introduced on the measurable space (E^*, B), where B is the Borel field generated by cylinder subsets of E^*. It also tells us that $\langle x, \xi \rangle, x \in E^*$, is a Gaussian random variable with mean 0 and variance $\|\xi\|_v^2$, where $\|\xi\|_v$ is the $L^2(R^d, v)$-norm.

Let $\langle x, \xi \rangle$, where $x \in E^*, \xi \in E$, be the canonical bilinear form, connecting E and E^*, defined for every x and ξ. Now x is viewed as a random parameter of a random variable $\langle x, \xi \rangle$ on the probability space (E^*, μ_v). On the other hand, this bilinear form is a continuous linear functional of ξ for fixed ξ and furthermore, it extends to a continuous linear functional of $f \in L^2(R^d, v)$. We denote it also by the form $\langle x, f \rangle$, although it is defined for almost all x. In view of these the $\langle x, f \rangle$ is called a *stochastic bilinear form.*

The collection $\{\langle x, f \rangle, f \in L^2(R^d)\}$ spans a subspace of $L^2(E^*, \mu_v)$ and is a Gaussian system denoted by $\mathbf{M}(\mu_v)$.

The following assertion can easily be proved, by noting that $E(|\langle x, f \rangle|^2) = \|f\|_v^2$ holds.

Proposition 3.2 *We have*

$$\mathbf{M}(\mu_v) \cong \mathbf{M}(Y) \cong L^2(R^d, v). \tag{3.1.6}$$

There is an observation and an important remark. If we assume that the measure v is the Lebesgue measure and $d = 1$, then, the characteristic functional, denoted by $C(\xi)$ is of the form

$$C(\xi) = \exp[-\frac{1}{2}\int \xi(t)^2 dt] = \exp[-\frac{1}{2}\|\xi\|^2].$$

Namely, we are given white noise. A realization is the time derivative of a Brownian motion, denoted by $\dot{B}(t)$. It is a generalized stochastic process which is *stationary* in time (the probability distribution μ is invariant under the time shift S_t : $S_t\dot{B}(s) = \dot{B}(s+t)$) and it has independent values at every instant in the sense of Gel'fand.

Stationarity enables us to speak of the *spectrum* in the following sense. Let (L^2) be the Hilbert space defined in Chapter 2. A member of (L^2) can be written in the form $\varphi(\dot{B}(s), s \in R)$. Then, the operator U_t defined by

$$(U_t\varphi)(\dot{B}(s), s \in R) = \varphi(\dot{B}(s+t), s \in R)$$

can be proved to be *unitary*. Obviously it satisfies the group property, i.e.

$$U_t U_s = U_{(t+s)}.$$

The continuity of U_t in t is enough to show

$$\text{s-}\lim_{t\to 0} U_t = I \quad \text{(identity)},$$

where s-lim means the strong limit. This comes from the fact that

$$\langle \dot{B}(\cdot + t), \xi\rangle = \langle \dot{B}(\cdot), S_t\xi\rangle$$

tends to $\langle \dot{B}(\cdot), \xi\rangle$ as $t \to 0$. Such a limit can be extended to Fourier-Hermite polynomials, and hence to the entire space.

Summing up, we are given a strongly continuous one–parameter group of unitary operators acting on (L^2). We can, therefore, appeal to Stone's theorem to have a spectral representation of U_t:

$$U_t = \int_{-\infty}^{\infty} e^{it\lambda} dE(\lambda),$$

where the system $\{E(\lambda); -\infty < \lambda < \infty\}$ is a resolution of the identity I, that is, $\{E(\lambda)\}$ is a system of projection operators such that

i) $E(\lambda)E(\mu) = E(\mu)E(\lambda) = E(\lambda \wedge \mu), \quad \lambda \wedge \mu = \min\{\lambda, \mu\}$,
ii) $\text{s-}\lim_{\lambda \to \mu+0} E(\lambda) = E(\mu)$,
iii) $E(\infty) = I$, and $E(-\infty) = 0$ (the zero operator).

The covariance function $\Gamma(s, t)$ of white noise is the delta function

$$\Gamma(t, s) = \delta(t - s).$$

On the other hand, Stone's theorem claims that $\dot{B}(t) = \int e^{it\lambda} dE(\lambda)\dot{B}(0)$. Hence, the covariance function is expressed in the form

$$\delta(t - s) = \int e^{i(t-s)\lambda} dm(\lambda),$$

where $dm(\lambda) = E[|E(\lambda)\dot{B}(0)|^2]$ which is $\frac{1}{2\pi} d\lambda$, since it is the Fourier transform of the delta function.

It has, so-to-speak a *flat spectrum* on the space H_1 as is seen from the above computation. This is the reason why it is called a white noise. It may be better to call Gaussian white noise, however as it is well known, we can just say white noise, unless confusion occurs, following the traditional way of naming.

As we shall see in next chapter, white noise is interpreted as *elemental* (or atomic).

We now come to another basic noise that is Poisson noise. Start with a Poisson process $P(t)$. For an interval $I = [a, b]$ we define $Y(I) = P(b) - P(a)$. This definition extends to the case of finite sum of disjoint intervals by additivity. Actually it is a sum of independent random variables. Then, Y is extended to the case where it is defined for an infinite sum of intervals; finally it is defined for any Borel set. As a result, we are given a random measure, which we call *Poisson random measure.*

Stochastic integrals based on Poisson random measure can be defined in a similar manner to the case of Gaussian case, and the same for stochastic integrals. We may use the notation $dP(t)$ or $\dot{P}(t)dt$ in this case, instead of

$dY(t)$ or $Y(dt)$. They are of the form

$$\int f(t)dP(t) \quad \text{or} \quad \int f(t)\dot{P}(t)dt.$$

Letting f run through the space $L^2(R)$, we can form a linear space that is formed by the stochastic integrals shown above.

The second formula using $\dot{P}(t)dt$ seems better so as to be understood as a linear functional of f, and it is even convenient to consider nonlinear functions of $\dot{P}(t)$'s.

As a system of elemental random variables, Gaussian white noise and Poisson noise are representative, however, from the viewpoint of innovation, one more case should be mentioned, namely *compound Poisson noise*. It appears in the decomposition of a Lévy process, but it is not atomic. It is a superposition of Poisson noises with various heights of jump. We shall deal with this case later, as it is not atomic.

3.2 Canonical representation of a Gaussian process

First we recall a definition of a Gaussian system. Let $\mathbf{X} = \{X_\alpha, \alpha \in A\}$ be a system of real random variables. If any finite linear combination of the X_α's is Gaussian in distribution, then the system $\mathbf{X}$ is called a *Gaussian system*. This definition is equivalent to the requirement that the joint distribution of any finite number of X_α's is Gaussian (may or may not be degenerated). The equivalence can easily be proved if we can actually use characteristic functions.

A Gaussian process $X(t), t \in T$, is a Gaussian system parametrized by $t \in T$, where T is either R^1 or its interval, say $[0, \infty)$ or $[a, b]$. The parameter t may be thought of as the time. Throughout this section, we always assume the following two conditions i) and ii).

i) $X(t)$ is separable.

Separability means that the subspace $M(X)$ of $L^2(\Omega, P)$ spanned by $X(t)$'s is a separable Hilbert space.

ii) $X(t)$ is purely non-deterministic (or has no remote past), namely

$$\bigcap_t M_t(X) = \{1\},$$

where $M_t(X)$ denotes the subspace of $M(X)$ spanned by the $X(s), s \leq t$ and $\{1\}$ stands for the trivial subspace involving only constants.

We can now state our main purpose of this section. Given a Gaussian process $X(t), t \in T$. We wish to have its representation in terms of a stochastic integral based on a Gaussian random measure chosen suitably. For this purpose, it is necessary to provide some background as is discussed below.

To fix the idea we take the parameter set T to be $[0, \infty)$. Our proposed representation of $X(t)$ is to be of the following form expressed as a stochastic integral

$$X(t) = \int_0^t F(t,u)Y(du), \ t \geq 0, \tag{3.2.1}$$

where $Y(B)$ is a Gaussian random measure with $E(Y(B)) = 0, E(Y(B)^2) = v(B)$, v being a Borel measure on $[0, \infty)$, and where $F(t,u), 0 \leq u \leq t$, is (t,u)-measurable and is square integrable in $u \in [0,t]$ for any t, i.e. a Volterra kernel.

Obviously, $E(X(t)) = 0$, and the covariance function $\Gamma(t,s)$ is

$$\Gamma(t,s) = \int_0^{t \wedge s} F(t,u)F(s,u)dv(u).$$

There arises a question if such a factorization of Γ always exists and further if such a representation in terms of F is unique in case it exists. The existence problem is a difficult question, so we shall discuss it later.

As for the uniqueness, we recall the famous Lévy's example which is surprising, before details are discussed.

Example 3.3 (P. Lévy) Take a white noise $\dot{B}(t)$ as a random measure.

Define $X_i(t), i = 1, 2$, by

$$X_1(t) = \int_0^t (2t - u)\dot{B}(u)du, \tag{3.2.2}$$

$$X_2(t) = \int_0^t (-3t + 4u)\dot{B}(u)du. \tag{3.2.3}$$

One can easily compute their covariance functions to confirm that both are the same; actual value is $\Gamma(t, s) = 3ts^2 - \frac{2}{3}s^3, \ s \le t$. This means that $X_1(t)$ and $X_2(t)$ are the same Gaussian process.

Here is a short note. A Gaussian process with mean 0 can uniquely be determined by the covariance function, since the probability distribution of the process is given by its finite dimensional joint distributions which are completely determined by covariance matrix. Hence, when we say that a cantered Gaussian process is given, we understand that a covariance function is given. The above example therefore tells that representation (3.2.1) of a given Gaussian process is not unique, although the equivalence of representations will be rigorously defined later.

Exercise. The reader is highly recommended to find any criterion which discriminates the above two representations before coming to Theorem 3.1. Also, find which is better in a sense of being significant.

P. Lévy proposed a particular representation called *canonical representation*, which has been known to be quite an important notion, not only for the representation theory of Gaussian processes, but also in general stochastic calculus, which will be discussed in Chapter 8.

Now we give the definition of canonical representation of a Gaussian process $X(t), t \ge 0$. In what follows, a Gaussian process $X(t)$ is always assumed to be centered; $E(X(t)) = 0$ for every t.

Definition 3.3 A representation of $X(t)$ given by the formula (3.2.1) is called *canonical* if the equation for the conditional expectation

$$E(X(t)|\mathbf{B}_s(X)) = \int_0^s F(t, u)Y(du)$$

holds for every $s \le t$, where $\mathbf{B}_s(X)$ is the smallest σ-field with respect to which all the $X(u), u \le s$, are measurable. The kernel function $F(t, u)$ is called a *canonical kernel.*

There is a sufficient condition for a representation to be canonical.

Proposition 3.3 *If $\mathbf{B}_s(X) = \mathbf{B}_s(Y)$ holds for every s, then the representation (3.2.1) is canonical.*

The proof is easy, and this condition is useful to see if a representation is canonical.

■

If the condition stated in Proposition 3.3 is satisfied, the representation is called *a proper canonical.* A canonical representation can easily be modified so as to be a proper canonical representation. Hence, in what follows, we shall omit "proper" and say simply "canonical". The same for the kernel of the represention, simply called "canonical kernel".

There is a criterion for a kernel to be canonical (see Hida[35]).

Theorem 3.1 *A representation (3.2.1) of $X(t)$ is canonical, if and only if, for any $t_0 \in T$,*

$$\int^t F(t,u)f(u)dv(u) = 0 \quad for\ every\ t \le t_0, \tag{3.2.4}$$

implies

$$f(u) = 0 \quad almost\ everywhere\ on\ (-\infty, t_0]\bigcap T. \tag{3.2.5}$$

Proof. If the representation is not canonical, then there exists a non-zero element $Z \in M_{t_0}(Y)$ which is independent of $X(t), t \le t_0$. Noting that Z is expressed in the form $\int^{t_0} f(u)Y(du), f \in L^2(v)$, we have the equation (3.2.5).

Conversely, if f satisfies (3.2.4) but not (3.2.5), then $M_t(X)$ is strictly less than $M_t(Y)$, so that the representation is not canonical.

■

Coming back to the Lévy's example, Example 3.3, the representation given by (3.2.2) is a canonical representation. But the representation given by (3.2.3) is *not* canonical, because we can choose $\int^{t_0} u^2\dot{B}(u)du$ to be the integral $\int^{t_0} f(u)Y(du)$ in the proof of Theorem 3.1. ($Y(du)$ is realized by $\dot{B}(u)du$.)

Equivalence of representations

A representation of a Gaussian process is determined by the pair $(F(t,u), Y)$. Since we are assuming that processes are Gaussian, so that we may take the pair $(F(t,u), v)$. Two representations $(F_i(t,u), v_i), i = 1,2$, are equivalent if

$$\int_C F_1(t,u)^2 v_1(du) = \int_C F_2(t,u)^2 v_2(du) \tag{3.2.6}$$

holds for every Borel subset C of T.

By this definition, we can immediately prove the following assertion due to P. Lévy since conditional expectation does not depend on the expressions of representation of the given process. The equality (3.2.6) asserts that the probability distribution of the system of conditional expectations is independent of the representations.

Proposition 3.4 *There exists at most one canonical representation.*

Example 3.4 Set $F(t,u) = f(u)\chi_{[0,t]}(u)$, where $f(u)$ is a step function taking values 1 or -1. Then

$$X(t) = \int_0^t F(t,u)\dot{B}(u)du$$

is a Brownian motion, since

$$E[X(t)X(s)] = E[\int_0^{t\wedge s} |f(u)|^2 du] = t \wedge s$$

and it is a canonical representation.

Other representation of a Brownian motion is

$$B(t) = \int_0^t 1\dot{B}(u)du.$$

The above two representations are equivalent since

$$\int_c |F(t,u)|^2 du = \int_C 1du$$

for any Borel set C.

Note that $\mathbf{B}_t(X) = \mathbf{B}_t(B)$ for every t, however sample functions behave differently.

As a consequence of Proposition 3.4, we can say that the structure of a Gaussian process can be characterized by the kernel (in fact, the canonical kernel) and the random measure (in fact the measure v). The kernel is not random and is known in the classical theory of functional analysis, while the random measure is easily studied since it is obtained from an additive Gaussian process. Thus, we have provided a pair of powerful tools, we are familiar to both.

Formally speaking, $Y(dt)$ or $dY(t)$ with $dt > 0$ is independent of $X(s), s \le t$, and it has the new information that $X(t)$ gains during the infinitesimal time interval $[t, t+dt)$. $Y(dt)$ is nothing but the *innovation* of $X(t)$ which we shall discuss in detail in a separate chapter, namely Chapter 8, as it is a very important concept. In fact the use of innovation reflects our idea of analysis. One may remind the word *reduction* appeared in Chapter 1, when we have tried to form the process as a functional of the innovation. Actually, we are in this line.

From the discussion above, one might think that the representation theory is rather elementary, however this is not quite correct. Profound structure exists behind the formula of canonical representations.

To make a concrete discussion, we now take a Brownian motion $B(t)$ to define a random measure, so that $Y = \dot{B}$ and a kernel function $G(t,u)$ is of Volterra type. Define a Gaussian process $X(t)$ by

$$X(t) = \int_0^t G(t,u)\dot{B}(u)du.$$

Now we assume that $G(t,u)$ is a smooth function on the domain $0 \le u \le t < \infty$ and $G(t,t)$ never vanishes.

Theorem 3.2 *The variation $\delta X(t)$ of the process $X(t)$ is well defined and is given by*

$$\delta X(t) = G(t,t)\dot{B}(t)dt + dt \int_0^t G_t(t,u)\dot{B}(u)du,$$

where $G_t(t,u) = \frac{\partial}{\partial t}G(t,u)$. $\dot{B}(t)$ is the innovation of $X(t)$ if and only if G(t,u) is a canonical kernel.

In the case of non-canonical representation, we have tacitly assumed that input signal $\dot{B}(t)$ is given in advance.

Remark 3.2 *In the variational equation, the two terms on the right-hand side seem to be of different order as dt tends to zero, so that two terms may be discriminated. But in reality the problem of discrimination of the terms is not so simple and not of our concern.*

Remark 3.3 *We often use the terms canonical representation and canonical kernel also for the second order stochastic process, which is a process with finite variance and is mean continuous. There the white noise is replaced by an orthogonal random measure. (cf. Innovation in the weak sense.)*

Now, one may ask when the canonical representation exists, or can it be constructed. To answer this question, we recall the Hellinger-Hahn theorem. Let T be a finite or infinite interval taken to be the time parameter set.

Theorem 3.3 The Hellinger-Hahn Theorem

Let $\{E(t), t \in T\}$ be a resolution of the identity I on a separable Hilbert space H. Then, there exists a system of members $\{f_n\}$ in H such that the H admits a direct sum decomposition of the form

$$H = \bigoplus_n H(f_n),$$

where $H(f_n)$ is a cyclic subspace spanned by $\{dE(t)f_n; t \in T\}$, with a cyclic vector f_n, and where

$$d\rho_{n+1}(t) << d\rho_n(t), \quad \rho_n(t) = \|E(t)f_n\|^2,$$

for every n. The item before $<<$ is absolutely continuous with respect to that after $<<$.

The direct sum decomposition is unique up to unitary equivalence, in particular the type of the measure $\{d\rho_n\}$ is independent of the choice of cyclic vectors $\{f_n\}$.

Definition 3.4 The number of non-trivial cyclic subspaces is called the *multiplicity* and the system $\{\rho_n, n \geq 1\}$ is the *spectral type*.

Remark 3.4 *What we discussed about spectrum of the shift operator in Section 3.1 is part of the Hellinger-Hahn theorem. There we have started from the one-parameter group of unitary operators, then applied Stone's theorem to obtain projections which form a resolution of the identity. The Hilbert space, on which the projections act, is H_1 spanned by linear functionals of white noise. This is the reason why we do not need to consider the case of higher multiplicity.*

In order to apply this theorem to the representation theory of a Gaussian process $X(t)$ with $E(X(t)) = 0,\ t \in T$, we shall proceed as follows:

1) Set $N_t(X) = \text{span}\{X(s), s \le t\}$ and set $M_t(X) = N_{t+}(X)$.
2) Set $M(X) = \bigvee M_t(X)$. Define a projection operator $E(t)$ such that

$$E(t) : M(X) \mapsto M_t(X).$$

3) Confirm the assumption that $X(t)$ is purely non-deterministic (or $X(t)$ has no remote past), namely

$$\bigwedge_t M_t = \{0\},$$

and $X(t)$ is separable, that is $M(X)$ is a separable Hilbert space. Then, we are ready to apply the Hellinger-Hahn theorem.
4) If the multiplicity of $X(t)$ is one, then the unique one–parameter family $E(t)f_1$ defines a Gaussian additive process, from which we can form the canonical representation of $X(t)$.

In a general case, the collection $\{dE(t)f_n, n = 1, 2, ...\}$ corresponds to the innovation of the $X(t)$ and defines a system of random measures. The representation is therefore a *generalized canonical representation* in the sense that higher multiplicity is involved. It is illustrated as follows.

Since we are concerned with a Gaussian system, it is better to write X_n instead of f_n. The $dE(t)X_n$ is a Gaussian random measure, so that it may be written as $Y(dt)$. Under the assumptions, $X(t)$ is therefore expressed in the form

$$X(t) = \sum_1^N \int^t F_n(t, u)dY(t), \tag{3.2.7}$$

where N is the multiplicity.

There naturally arises a question: Given a Gaussian process, how to construct a canonical or a generalized canonical representation.

Before we discuss this problem, definition of a stochastic process should well be understood to be a consistent family of joint distributions, and we specify the story to a Gaussian process. Assuming that the expectation (which is always assumed to be 0) and the covariance function are given. For a Gaussian process, the covariance function determines all the joint distributions, so that a Gaussian process is given uniquely by specifying the covariance function.

With a covariance function, the associated reproducing kernel Hilbert space is given. (See Chapter 2. A.3.) There we have projections $E(t)$ and apply the Hellinger-Hahn theorem. The cyclic subspaces and kernels are to be known theoretically, so that we have a generalized canonical representation. The multiplicity is determined at this stage. Random version from the result in RKHS can also be formed if one wishes.

Remark 3.5 *In what follows we usually consider canonical representations in the case where T is a finite or infinite interval and the spectral measure $d\rho_1$ is taken to be the Lebesgue measure on T.*

Various examples of a representation are now in order.

Example 3.5 Define $X(t)$ by stochastic integral based on $\dot{B}(t)$:

$$X(t) = \int_0^t (t-u)\dot{B}(u)du,$$

which gives a canonical representation. Its derivative (in the $L^2(\Omega, P)$-sense) is a Brownian motion.

Example 3.6 Consider a Gaussian process given by

$$X(t) = \int_0^t (2 - \frac{3u}{t})\dot{B}(u)du$$

which is proved to be a Brownian motion by computing the covariance function. It is, in fact, a non-canonical, therefore we have a non-trivial representation of a Brownian motion by using a non-canonical kernel.

Example 3.7 (M. Hitsuda[73]) Let $\mathbf{B}(t) = (B_1(t), B_2(t), \cdots, B_n(t))$ be an n-dimensional Brownian motion, i.e. B_j's are mutually independent Brownian motions. We provide a function $F(t)$ such that

i) it is locally bounded,
ii) it is absolutely continuous,
iii) its Radon-Nikodym derivative belongs to $L^1([a,b])$, but not to $L^2([a,b])$ for any interval $[a,b]$.

Then, define $X(t), t \geq 0$, by

$$X(t) = \sum_{k=1}^{n} F^{(k-1)}(t) B_k(t),$$

where $F^0(t) = F(t)$, $F^{(k)}(t)$ is the k-th derivative of $F(t)$ and where $B_k(t)$'s are independent Brownian motion.

It is proved (see the literature[73]) that $X(t)$ has multiplicity n. The above equation shows that we are given a generalized canonical representation.

Example 3.8 Let $B_i(t), i = 1, 2$, be two Brownian motions which are mutually independent. Define a Gaussian process $X(t), t \geq 0$, by

$$X(t) = \begin{cases} B_1(t), & 0 \leq t \leq 1, \\ B_1(t), & t \geq 1, \text{rational}, \\ B_2(t), & t \geq 1, \text{irrational}. \end{cases}$$

Then, it is proved that the $X(t)$ has double multiplicity and that the spectral measure $d\rho_1$ is of Lebesgue type on $[0, \infty)$, and $d\rho_2$ is of Lebesgue type, too, but supported by $[1, \infty)$.

It is easy to write the generalized canonical representation of $X(t)$. It is expressed in the form

$$X(t) = \int_0^t F_1(t,u) \dot{B}_1(u) du + \int_0^t F_2(t,u) \dot{B}_2(u) du,$$

where $F_1(t,u)$ and $F_2(t,u)$ are as follows.

For $t \leq 1$,

$$F_1(t,u) = 1,$$
$$F_2(t,u) = 0,$$

and for $t \geq 1$,

$$F_1(t, u) = \begin{cases} \chi_{[0,t]}(u), & t \text{ is rational,} \\ 0, & t \text{ is irrational,} \end{cases}$$

$$F_2(t, u) = 1 - F_1(t, u).$$

It is possible to form a single Brownian motion $B(t)$ to have a representation

$$Y(t) = \int_0^t F(t, u)\dot{B}(u),$$

where $Y(t)$ has the same probability distribution as $X(t)$. The above representation is, of course, non-canonical.

It may be interesting to show a concrete construction of a non-canonical representation of $X(t)$ given in Example 3.8 that combines two Brownian motions. Actual random measure and kernel are as follows.

A new white noise $\dot{B}(t)$ is defined by

$$\dot{B}(t) = \begin{cases} \frac{1}{\sqrt{2}}\dot{B}_1(2t - n), & t \in [n, n + \frac{1}{2}], \\ \frac{1}{\sqrt{2}}\dot{B}_2(2t - n), & t \in [n + \frac{1}{2}, n + 1], n \geq 0. \end{cases}$$

The kernel $F(t, u)$ is

i) For $0 \leq t \leq \frac{1}{2}$,

$$F(t, u) = 1, \quad t \in [0, \frac{1}{2}].$$

ii) For $t \geq \frac{1}{2}$, we have,

a) if $t \in [n + \frac{1}{2}, n + 1]$, with $n \geq 0$,

$$F(t, u) = \begin{cases} 0, & t \text{ is rational} \\ 1, & t \text{ is irrational} \end{cases}$$

b) if $t \in [n, n + \frac{1}{2}]$, $\quad n \geq 1$,

$$F(t, u) = \begin{cases} 1, & t \text{ is rational} \\ 0, & t \text{ is irrational.} \end{cases}$$

Note that

$$M_t(X) = M_t(B_1) \bigoplus \{M_t(B_2) \bigcap M_t(B_2)^{\perp}\},$$

while

$$M_t(X) \subset M_t(B).$$

It is worthwhile to mention that the *time operator* defined below for a stationary process will be helpful as a second operator to investigate the evolution and multiplicity.

Definition 3.5 Set

$$T = \int t dE(t),$$

then T is a self-adjoint operator and is called the *time operator*.

The time operator will be discussed in Chapter 10, Section 10.4.

Assume that $X(t)$ has unit multiplicity, that is, it has the canonical representation. Having obtained the innovation, which may now be assumed to be a white noise $\dot{B}(t)$, we can define the partial derivative denoted by ∂_t with respect to $\dot{B}(t)$, since it is elemental and is viewed as a variable. Functionals of the $X(t)$'s are also functionals (now only linear functionals are concerned) of the $\dot{B}(t)$'s. Hence ∂_t can be applied to members in $M(X)$.

With this situation the canonical kernel can be obtained by

$$\partial_u X(t) = F(t, u), \quad u < t.$$

Generalization of this fact to the case with higher multiplicity can be given, although expressions will be complex.

3.3 Multiple Markov Gaussian processes

Let $X(t), t \in R$, be a Gaussian process with canonical representation

$$X(t) = \int_{-\infty}^{t} F(t, u) \dot{B}(u) du, \tag{3.3.1}$$

where the white noise is the innovation of $X(t)$.

We keep the notations such as $M_t(X)$, $M(X)$, $\mathbf{B}_t(X)$ and others that were established in the last section. The multiple Markov property of $X(t)$ was defined in Hida[35] in 1960. However, to make this note self-contained, we repeat the definition. As usual it is assumed that $E(X(t)) = 0$ and $X(t)$ is separable.

Definition 3.6 If the conditional expectations $\{E(X(t_i)|\mathbf{B}_{t_0}(X)); i = 1, 2, \cdots, N\}$ are linearly independent for any t_i's with $t_0 \le t_1 < t_2 < \cdots < t_N$, and if $\{E(X(t_i)|\mathbf{B}_{t_0}(X)); i = 1, 2, \cdots, N+1\}$ are linearly dependent for any t_i's with $t_1 < t_2 < \cdots < t_{N+1}$, then $X(t)$ is called an N-ple Markov process.

Theorem 3.4 *Let $X(t)$ be an N-ple Markov Gaussian process with $E(X(t)) = 0$. Assume that $X(t)$ has a canonical representation and that $M_t(X)$ is continuous in t. Then, its canonical kernel $F(t,u)$ is a Goursat kernel of order N.*

A kernel $F(t,u), u \le t$, is a Goursat kernel, if it is expressed in the form

$$F(t,u) = \sum_{i=1}^{N} f_i(t) g_i(u), \tag{3.3.2}$$

and if the matrix $(f_i(t_j))$ for different t_j's is nonsingular and g_j's are linearly independent in $L^2([0,t])$ for every t.

For proof we refer to Hida[35] and Hida-Hitsuda[59].

A 1-ple Markov Gaussian process $X(t), t \ge 0$, is a simple Markov process in the ordinary sense. It is expressed in the form

$$X(t) = f(t) \int_0^t g(u) \dot{B}(u) du = f(t) U(t), \tag{3.3.3}$$

where $g(t)$ never vanishes. Note that $U(t)$ is an additive process.

The following example comes from Petz's example as a continuous version.

Example 3.9 (D. Petz) Suppose a 1-ple Markov Gaussian process $X(t), t \geq 0$, is expressed in the form (3.3.3). Then, the following equalities hold:

$$\begin{aligned} E(X(t_1)X(t_2)^2X(t_3)) &= 3E(X(t_1)X(t_3))E(X(t_2)^2) \\ &= 3E(X(t_1)X(t_2))E(X(t_2)X(t_3)), \end{aligned}$$

where $(t_2 - t_1)(t_3 - t_2) > 0$ is assumed.

Proof. Assume the inequality $t_1 < t_2 < t_3$ holds. Use the formula (3.3.3) to compute

$$E(X(t_1)X(t_2)^2X(t_3)) = f(t_1)f(t_2)^2f(t_3)E(U(t_1)U(t_2)^2U(t_3)) \quad (3.3.4)$$

$$= f(t_1)f(t_2)^2f(t_3)E((U(t_1)U(t_2)^3))$$

$$= f(t_1)f(t_2)^2f(t_3)E((U(t_1)((U(t_1) + U(t_2) - U(t_1))^3)$$

$$= f(t_1)f(t_2)^2f(t_3)3\left\{\left(\int_0^{t_1} g(u)^2du\right)^2 + \int_0^{t_1} g(u)^2du \int_{t_1}^{t_2} g(u)^2du\right\}.$$

In a similar manner we can compute $3E(X(t_1)X(t_3))E(X(t_2)^2)$ and $3E(X(t_1)X(t_2))E(X(t_2)X(t_3))$, which are equal to the above result.

The $X(t)$ in Example 3.3 is a 2-ple (double) Markov Gaussian process.

The following theorem is well known.

Theorem 3.5 *Suppose that $X(t)$ is a purely non-deterministic, mean continuous and stationary Gaussian process. Then, it has a canonical representation and the canonical kernel $F(t,u)$ is a function only of $t-u$. Namely,*

$$F(t,u) = F(t-u), \quad u \leq t.$$

The next lemma should be noted.

Lemma 3.1 *Let $F(t,u)$ be a Goursat kernel of order N expressed in the form (3.3.2). If $F(t,u)$ is a function only of $(t-u)$, then $\{f_i(t)\}$ is a fundamental system of solutions of ordinary differential equation of order*

N with constant coefficients, and $\{g_i(u)\}$ is also a fundamental system of solutions of the adjoint differential operator.

The proof of this lemma can also be seen in the literatures[35,59].

The explicit forms of the terms of $F(t-u)$ are listed below:

$$t^k u^{n-k} e^{-\lambda(t-u)} \sin \mu(t-u), \tag{3.3.5}$$

$$t^k u^{n-k} e^{-\lambda(t-u)} \cos \mu(t-u), \tag{3.3.6}$$

where $u \leq t, 0 \leq k \leq n, n \leq N$, μ real.

Note that *sine* and *cosine* are used to avoid the complex valued solutions to the characteristic equation of the ordinary differential equation in question.

So far we have considered the stationary processes, that is, the probability distribution of the process is invariant under the time shift. Contrary to this, we may consider the *dilation* invariance of a process. It means that for a stochastic process $X(t), t \geq 0$, we claim

$$X(at) \quad \sim \quad \sqrt{a}X(t), \quad a > 0,$$

where $\sim$ means both sides have the same probability distribution.

Theorem 3.6 *Let $X(t), t \geq 0$, have a canonical representation and be dilation invariant. Then its canonical kernel $F(t,u)$ is a function of u/t.*

Proof. By assumption, we have

$$\int_0^{at} F(at,u)\dot{B}(u)du \ \sim \ \sqrt{a}\int_0^t F(t,u)\dot{B}(u)du.$$

Since $\dot{B}(av)$ has the same distribution as $\frac{1}{\sqrt{a}}\dot{B}(v)$, the left-hand side of the above equation is subject to the same distribution as $\sqrt{a}\int_0^t F(at,au)\dot{B}(u)du$. This property holds for every t and a. The uniqueness of the canonical kernel (up to sign) implies that we have

$$F(at,au) = F(t,u), \ \ a.e.\ u \leq t. \tag{3.3.7}$$

This means the invariance of $F(t, u)$ holds under the *isotropic* dilation:

$$(t, u) \mapsto (at, au), \ a > 0. \tag{3.3.8}$$

This implies

$$F(\frac{u}{t}, 1) = F(t, u),$$

which directs us to the conclusion.

■

Lemma 3.2 *Let $F(t, u) = \sum_1^N f_i(t) g_i(u)$ be a Goursat kernel of order N. If $F(t, u)$ is dilation invariant, i.e. invariant under the dilation (3.3.8), then $\{f_i\}$ is a fundamental system of solutions to Euler equation of order N of the form*

$$t^N f^{(N)}(t) + a_1 t^{(N-1)} f^{(N-1)}(t) + \cdots + a_N f(t) = 0,$$

while $\{g_i\}$ is a fundamental system of solutions to the adjoint equation.

This lemma comes from Lemma 3.1, by changing the variables

$$t \longrightarrow e^t, \quad u \longrightarrow e^u.$$

If imaginary roots of the characteristic equation are ignored, we have

$$F(t, u) = \sum_{j,k} a_{j,k} \left(\log \frac{u}{t}\right)^j \left(\frac{u}{t}\right)^{\alpha_k},$$

where j, k are non-negative integers and are chosen so as the sum involves as many as N terms.

Lévy's Brownian motion, which is the most important random field, gives us interesting examples of multiple Markov Gaussian process. In fact, interesting properties of those multiple Markov processes help us to study dependence on the parameter of the Lévy's Brownian motion.

We give the definition.

Definition 3.7 A Gaussian system $\{B(a), a \in R^d\}$ is called *Lévy's Brownian motion* if it satisfies the following conditions:

i) $B(o) = 0, o$ is the origin of R^d,

ii) $E(B(a)) = 0, a \in R^d$ and

iii) $E(|B(a) - B(b)|^2 = |b - a|$.

If the parameter is restricted to a straight line passing through the origin, then we are given the ordinary Brownian motion.

Suppose d is odd, say $d = 2p + 1$. Then, we can prove that the $M_d(t)$ process, which is defined to be the average of the d-dimensional parameter Lévy Brownian motion over a sphere with center o and radius t, has the canonical representation, the canonical kernel of which is a polynomial in $\frac{u}{t}$ of degree $p + 1$. The polynomial has some characteristic properties. The property can be clarified in explicit forms of the kernel, if the $M_d(t)$ is transformed to a stationary process $Y(t)$ by the change of time

$$Y(t) = e^{-t} M(e^{2t}).$$

See P. Lévy[105], M. P. McKean[116].

3.4 Fractional Brownian motion

We are going to discuss a class of Gaussian processes that do not satisfy any Markov properties, but satisfy self-similarity. They can be studied in connection with the canonical representation theory. We know that such processes have practical interest.

It is noted that those processes were introduced much earlier. For instance we refer to the paper by A. N. Kolmogorov [85].

Following P. Lévy's book[104] Section 3.4 and B. Mandelbrot[111], we consider a Gaussian process $X_\alpha(t), t \geq 0$, given by the formula (we take the parameter $\alpha - 1$ instead of α for convenience).

$$X_\alpha(t) = \frac{1}{\Gamma(\alpha)} \int_0^t (t-u)^{\alpha-1} \dot{B}(u) du, \quad \alpha > \frac{1}{2}. \tag{3.4.1}$$

We have $E(X_\alpha(t)) = 0$, and $E(X_\alpha(t)^2) = \frac{t^{2\alpha-1}}{\Gamma(\alpha)^2(2\alpha-1)}$.

If we use the notation I_α that defines the fractional order integration operator

$$(I_\alpha f)(t) = \frac{1}{\Gamma(\alpha)} \int_0^t (t-u)^{\alpha-1} f(u) du \tag{3.4.2}$$

for a locally square integrable function f on $[0, \infty)$ (notation $f \in L^2_{loc}$), then $X_\alpha(t)$ may be formally expressed in the form

$$X_\alpha(t) = (I_\alpha \dot{B})(t), \tag{3.4.3}$$

although $\dot{B}(t)$ is not ordinary square integrable function. The above expression, however, makes sense if the integral is understood as a stochastic integral.

With this notation $\{(I_\alpha \dot{B})(t), t \geq 0, \alpha > \frac{1}{2}\}$ is a one-parameter family of Gaussian processes. In particular

$$(I_0 \dot{B})(t) = B(t).$$

It is noted that I_α acting on white noise $\dot{B}(t)$ is a *causal* operator in the sense that

$$\mathbf{B}_t(X_\alpha) \subset \mathbf{B}_t(\dot{B})$$

holds for every t.

One may then ask if equality holds in the above formula. In other words, the expression of $X_\alpha(t)$ is a canonical representation or not.

Before answering this question, we give some background in the following. The domain of I_α is taken to be the class L^2_{loc}. The following evaluations of the integral prove that for $f \in L^2_{loc}$, the image $I_\alpha f$ is again in L^2_{loc}. For any fixed $T > 0$, and for $f \in L^2_{loc}$

$$\begin{aligned}
\int_0^T (I_\alpha f)(t)^2 dt &= \frac{1}{\Gamma(\alpha)^2} \int_0^T \left(\int_0^T (t-u)^{\alpha-1} f(u) du \right)^2 dt \\
&\leq \frac{1}{\Gamma(\alpha)^2} \int_0^T \left(\int_0^t (t-u)^{2\alpha-2} du \int_0^t f(u)^2 du \right) dt \\
&= \frac{1}{\Gamma(\alpha)^2 (2\alpha-1)} \int_0^T \left(t^{2\alpha-1} \int_0^t f(u)^2 du \right) dt \\
&\leq \frac{T^{2\alpha}}{\Gamma(\alpha)^2 (2\alpha-1)} \int_0^T f(u)^2 du < \infty.
\end{aligned}$$

Thus, iterations of the I_α's can also be defined. With a short modification we can prove that this fact is true even in the case of stochastic integrals.

Proposition 3.5 *The fractional order integral operator I_α, given by (3.4.2) satisfies*

$$I_\alpha I_\beta = I_{\alpha+\beta}, \quad \alpha, \beta > \frac{1}{2},$$

where I_1 is just the integration.

It is easy to see that $(I_\alpha f)(t)$ is differentiable if $\alpha > 3/2$. In the case of $X_\alpha(t)$, it is differentiable in the mean square norm for $\alpha > 3/2$ and we obtain

$$\frac{d}{dt} X_\alpha = X_{\alpha-1}(t).$$

More generally, for $\alpha > 3/2$, set $k = [\alpha - \frac{1}{2}]$. Then, $X_\alpha(t)$ is k-times differentiable and we have

$$\frac{d^k}{dt^k} X_\alpha(t) = X_{\alpha-k}(t).$$

The notation $[\alpha]$ denotes the largest integer that is less than or equal to α.

It is easy to see that

$$I_\alpha X_\beta(t) = X_{\alpha+\beta}(t).$$

We now prove the theorem

Theorem 3.7 *The representation of Gaussian process $X_\alpha(t)$ expressed as a stochastic integral*

$$X_\alpha(t) = \frac{1}{\Gamma(\alpha)} \int_0^t (t-u)^{\alpha-1} \dot{B}(u) du \tag{3.4.4}$$

is canonical.

Proof. Set $k = [\alpha - \frac{1}{2}]$ (which is the largest integer $\leq \alpha - \frac{1}{2}$), and obtain $\frac{d^k}{dt^k} X_\alpha(t)$ which is written as $X_\beta(t)$ with $\beta = \alpha - k$. Now $\frac{1}{2} < \beta < \frac{3}{2}$ holds. We can therefore apply $I_{2-\beta}$ to $X_\beta(t)$ to obtain a process which may be expressed as $X_2(t)$. Obviously $X_2(t) = \int (t-u)\dot{B}(u)du$ is differentiable and its derivative is a Brownian motion. We therefore have inequalities

$$\mathbf{B}_t(B) \subset \mathbf{B}(X_2) \subset \mathbf{B}(X_\alpha).$$

Since

$$\mathbf{B}(X_\alpha) \subset \mathbf{B}_t(B)$$

is obvious by definition. Hence all the inclusions above are replaced by *equality*. This fact proves the theorem.

■

It is recalled that if α is a positive integer, say N, then $X_\alpha(t)$ is N-times differentiable, and the $\dot{B}(t)$, which is an innovation, can be obtained an N-th order differential operator, that is a local operator. This implies the multiple Markov property.

Remark 3.6 *In fact, $X_\alpha(t), \alpha = N$, is N-ple Markov process in the restricted sense. For details see the paper*[35].

On the other hand, if α is not an integer, then the operator to get innovation is not a local operator, but it is causal.

Self-similarity of $X_\alpha(t), t \geq 0$, can be stated as follows.

Theorem 3.8 *For any positive h and τ, the probability distribution of the increment*

$$X_\alpha(t + h\tau) - X_\alpha(t), \ t \geq 0,$$

is the same as that of

$$h^{-\alpha+\frac{1}{2}}\{X_\alpha(t+\tau) - X_\alpha(t)\}, \ t \geq 0.$$

Namely, self-similarity of order $h^{-\alpha+\frac{1}{2}}$ holds.

Proof. It is known that for any $h > 0$, $\{\dot{B}(ht)\}$ and $\{h^{-\frac{1}{2}}\dot{B}(t)\}$ have the same probability distribution. By using this property, the change of probability distributions of stochastic integrals according to the dilation of time $\tau \to h\tau$ can be evaluated and seen to be the same. Hence the theorem is proved.

■

Somewhat different considerations on variations of $X_\alpha(t)$ are made as follows. Actually we study three cases.

Variation 1. Parameter space R^1.

Since the kernel function of the canonical representation of $X_\alpha(t)$ is a function of $(t-u)$, we may think of stationarity of the process. From this viewpoint one will remind the formula given in the paper by B. Mandelbrot and J. Van Ness[111]. In their terminology, H (with $0 < H < 1$) corresponds to $\alpha - \frac{1}{2}$ in our notation. They define a *reduced fractional Brownian motion* $B_H(t)$ with parameter H as follows.

For $t = 0$,

$$B_H(0) = b_0,$$

and for $t > 0$

$$\begin{aligned} B_H(t) - B_H(0) = & \frac{1}{\Gamma(H+\frac{1}{2})}\left\{\int_{-\infty}^{0} [(t-u)^{H-\frac{1}{2}} - (-u)^{H-\frac{1}{2}}\}\dot{B}(u)du\right\} \\ & + \frac{1}{\Gamma(H+\frac{1}{2})}\left\{\int_{0}^{t} (t-u)^{H-\frac{1}{2}}\dot{B}(u)du\right\}. \end{aligned}$$

Self-similarity of $B_H(t), t \in R$, can be proved.

Variation 2. Some stationary processes

For the case $\alpha = 1$ we have a Brownian motion. By the change of time together with multiplication factor

$$B(t) \;\to\; e^{-t}B(e^{2t}) = U(t),$$

we are given a stationary Gaussian process, an Ornstein-Uhlebeck process with spectral density function $\frac{1}{(1+\lambda^2)}$ up to constant. A generalization is as follows: If α is an integer, say N, then a transformation, a generalization of the above time change, is

$$X_\alpha(t) \;\to\; e^{-t}X_N(e^{2t}) = U_N(t)$$

which is a stationary Gaussian process with spectral density function

$$f_N(\lambda) = c_N \frac{1}{(1+\lambda^2)^N}.$$

The multiple Markov property is inherited by such a time change.

Now, as a generalization of this, we may consider a stationary Gaussian process $U_\alpha(t)$ with spectral density function

$$f_\alpha(\lambda) = c_\alpha \frac{1}{(1+\lambda^2)^\alpha}.$$

The process $U_\alpha(t)$ may be called stationary α-ple Markov Gaussian process.

We can easily prove that it has a canonical representation and that it satisfies self-similarity.

Variation 3. Complexification

Let $Z(t), t \geq 0$, be a complex Brownian motion. Define

$$X_\alpha(t) = (I_\alpha \dot{Z})(t) = \frac{1}{\Gamma(\alpha)} \int_0^t (t-u)^{\alpha-1} \dot{Z}(u) du. \qquad (3.4.5)$$

The $Z_\alpha(t)$ defined above is a *complex fractional Brownian motion.*

This has been defined by P. Lévy [104], Section 3.4. The stochastic integral defining the $X_\alpha(t)$ is well defined in any finite interval provided $\alpha > -1/2$. P. Lévy also noticed that, for $0 > \alpha > -1/2$, this corresponds to the fractional derivative of order $-\alpha$, in the sense of Riemann, of the Brownian motion.

We consider this fact in some more details. Since the operator I_α is invertible because it can be reduced, up to a shift in the argument, to a standard Volterra operator. Hence, the complex white noise $\dot{Z}(t)$ can be expressed as a function of $X_\alpha(t)$:

$$\dot{Z}(u) = I_\alpha^{-1}(X_\alpha(u))$$

and hence we see that the representation (3.4.4) is canonical. We may be allowed to speak of canonical representation for complex Gaussian process without giving definition.

3.5 Stationarity of fractional Brownian motion

Following the method used in the previous section we would like to discuss modification to make the fractional Brownian motion stationary. First of all to make the process (3.4.5) stationary, one has to take the time interval to be the entire R^1. Since the function $(t-\tau)^\alpha$, for $\alpha > -1/2$, is not square integrable in $(-\infty, t]$, we need some kind of renormalization.

The problem is now to modify the stochastic integral

$$\int_{-\infty}^{t} \frac{(t-\tau)^\alpha}{\Gamma(\alpha+1)} \dot{B}(u)du. \tag{3.5.1}$$

It formally defines a stationary (Gaussian) process, but the stochastic integral does not converge. Suppose we can find a function $\gamma(\tau)$ such that, for every $t \in \mathbf{R}$ the integral

$$\int_{-\infty}^{t} \frac{(t-\tau)^\alpha}{\Gamma(\alpha+1)} \dot{B}(\tau)d\tau - \int_{-\infty}^{0} \gamma(\tau)\dot{B}(\tau)d\tau \tag{3.5.2}$$

is well defined. Then, we can expect that the resulting process is stationary for the time shift of the white noise. This conjecture is based on the following heuristic argument.

Let T_s^B denote the shift of the white noise $\dot{B}$. Applying T_s^B to the formal expression (3.5.2) we obtain the new process (indexed by $t \in \mathbb{R}_+$)

$$\int_{-\infty}^{t} \frac{(t-\tau)^\alpha}{\Gamma(\alpha+1)} \dot{B}(\tau+s)d\tau - \int_{-\infty}^{0} \gamma(\tau)\dot{B}(\tau+s)d\tau. \tag{3.5.3}$$

Since the process $\{\dot{B}(\tau+s)\}$ is equivalent in law to the process $\{\dot{B}(\tau)\}$, it follows that the process expressed in (3.5.3) is equal in law to the process defined by (3.5.2).

In what follows we will show that the above heuristic considerations lead us to a correct result.

Lemma 3.3 *For any $t \in R$ the function*

$$F(t-\tau) = (t-\tau)^{2\alpha}, \ \tau \le t,$$

is locally integrable in $(-\infty, t]$ if $\alpha > -\frac{1}{2}$ and is not locally integrable in $(-\infty, t]$ if $\alpha \le -\frac{1}{2}$.

Proof. For ε, $T > 0$ one has, by elementary computations,

$$\int_{-T}^{\varepsilon} (t-\tau)^{2\alpha} d\tau = -\int_{t+T}^{t-\varepsilon} \sigma^{2\alpha} d\sigma = \int_{t-\varepsilon}^{t+T} \sigma^{2\alpha} d\sigma$$

$$= \begin{cases} \lg \sigma \Big|_{t-\varepsilon}^{t+T}, & \text{if} \quad \alpha = -1/2 \\ \frac{\sigma^{2\alpha+1}}{2\alpha+1} \Big|_{t-\varepsilon}^{t+T}, & \text{if} \quad \alpha \neq -1/2. \end{cases}$$

Thus, if $\alpha \leq -1/2$, the integral has a singularity at zero and the function is not locally integrable. ■

Lemma 3.4 *For any $t \in \mathbf{R}_+$ and for $-1/2 < \alpha < 1/2$ the function $G(t,\tau)$ given by*

$$G(t,\tau) = (t-\tau)^\alpha - (-\tau)^\alpha$$

is square-integrable over $(-\infty, t]$

Proof. To prove the square integrability of $G(t,\tau)$ in $(-\infty, t]$ we need only to study its behavior as $\tau \to -\infty$. In this limit one has

$$[(t-\tau)^\alpha - (-\tau)^\alpha]^2 = \tau^{2\alpha}\left[\left(\frac{t}{\tau} - 1\right)^\alpha - 1\right]^2 \approx \tau^{2\alpha}\left[\frac{t}{\tau}\right]^2 \approx \tau^{2(\alpha-1)},$$

for fixed t. Therefore, the tail part of the integral behaves like

$$\tau^{2(\alpha-1)+1},$$

hence the integral is finite for

$$2(\alpha - 1) + 1 < 0 \Leftrightarrow \alpha < \frac{1}{2}.$$

■

Theorem 3.9 *Let $\alpha \in (-\frac{1}{2}, \frac{1}{2})$ and let $X_\alpha(0)$ be a Gaussian random variable defined by*

$$X_\alpha(0) = \int_{-T}^{0} \frac{(-\tau)^\alpha}{\Gamma(\alpha+1)} \dot{B}(\tau) d\tau.$$

Then the limit

$$X_\alpha(0) + \lim_{T\to+\infty} \left\{ \int_{-T}^{t} \frac{(t-\tau)^\alpha}{\Gamma(\alpha+1)} \dot{B}(\tau) d\tau - \int_{-T}^{0} \frac{(-\tau)^\alpha}{\Gamma(\alpha+1)} \dot{B}(\tau) d\tau \right\} \quad (3.5.4)$$

exists in (L^2). Let it be denoted by $X_\alpha(t)$. It is a Gaussian process.

Proof. Let

$$X(t) = \int_{-\infty}^{t} \frac{(t-u)^\alpha}{\Gamma(\alpha+1)} \dot{B}(u)du - \int_{-\infty}^{0} \frac{(-u)^\alpha}{\Gamma(\alpha+1)} \dot{B}(u)du$$
$$= \int_{0}^{t} \frac{(t-\tau)^\alpha}{\Gamma(\alpha+1)} \dot{B}(u)du + \int_{-\infty}^{0} \frac{(t-\tau)^\alpha - (-\tau)^\alpha}{\Gamma(\alpha+1)} \dot{B}(u)du. \quad (3.5.5)$$

Since $\alpha \in (-1/2, 1/2)$, the function $(t-\tau)^\alpha - (-\tau)^\alpha$ is square integrable in $(-\infty, t]$ by Lemma 3.4. Therefore the sum

$$X_\alpha(0) + X(t)$$

is defined to be a member in (L^2). Moreover, it is the same process as $X_\alpha(t)$ given by (3.5.4).

■

Definition 3.8 The process $\{X_\alpha(t)\}$, defined by Theorem 3.9 is called a *stationary fractional Brownian motion* of order $\alpha \in (-\frac{1}{2}, \frac{1}{2})$.

Set

$$X_T(t) = \int_{-T}^{t} \frac{(t-u)^\alpha}{\Gamma(\alpha+1)} \dot{B}(u)du - \int_{-T}^{0} \frac{(-u)^\alpha}{\Gamma(\alpha+1)} \dot{B}(u)du. \quad (3.5.6)$$

Define a translation $\mathbf{T}_s^t$, for each $s \in \mathbb{R}_+$, such that

$$\mathbf{T}_s^t X_T(t) := \int_{-T}^{t+s} \frac{(t+s-u)^\alpha}{\Gamma(\alpha+1)} \dot{B}(u)du - \int_{-T}^{s} \frac{(s-u)^\alpha}{\Gamma(\alpha+1)} \dot{B}(u)du. \quad (3.5.7)$$

Proposition 3.6 *The limit*

$$\lim_{T\to+\infty} \mathbf{T}_s^t X_T(t) \quad (3.5.8)$$

exists in (L^2) and is equal, in law, to the process $(X = X(t))$ where $X(t)$ is given in (3.5.5).

Proof. The process $\mathbf{T}_s^t X_t(t)$ is well defined for each T, s, $t \geq 0$ because of Lemmas 3.3 and 3.4. With the change of variables $\tau = u - s$, $T_s^t X(t)$ becomes

$$\int_{-(T+s)}^{t} \frac{(t-\tau)^\alpha}{\Gamma(\alpha+1)} \dot{B}(\tau+s)d\tau - \int_{-(T+s)}^{0} \frac{(-\tau)^\alpha}{\Gamma(\alpha+1)} \dot{B}(\tau+s)d\tau.$$

But the stochastic integral

$$\int_{-\infty}^{0} \left[\frac{(t-\tau)^\alpha}{\Gamma(\alpha+1)} - \frac{(-\tau)^\alpha}{\Gamma(\alpha+1)}\right] \dot{B}(\tau+s)d\tau$$

is well defined for all s because of Lemma 3.4, therefore the limit (3.5.8) exists and is equal to

$$\int_{0}^{t} \frac{(t-\tau)^\alpha}{\Gamma(\alpha+1)} \dot{B}(\tau+s)d\tau + \int_{-\infty}^{0} \left[\frac{(t-\tau)^\alpha}{\Gamma(\alpha+1)} - \frac{(-\tau)^\alpha}{\Gamma(\alpha+1)}\right] \dot{B}(\tau+s)d\tau.$$

By the stationarity of the white noise, this integral is equal, in law, to the process

$$\int_{0}^{t} \frac{(t-\tau)^\alpha}{\Gamma(\alpha+1)} \dot{B}(\tau)d\tau + \int_{-\infty}^{0} \left[\frac{(t-\tau)^\alpha}{\Gamma(\alpha+1)} - \frac{(-\tau)^\alpha}{\Gamma(\alpha+1)}\right] \dot{B}(\tau)d\tau = X(t)$$

and this proves the statement of Proposition 3.6.

■

3.6 Fractional order differential operator in connection with Lévy's Brownian motion

The R^2 parameter Lévy Brownian motion

As is well-known, the collection $\left\{\frac{1}{\sqrt{2\pi}}; \sqrt{\pi}\cos k\theta, \sqrt{\pi}\sin k\theta; k \geq 1\right\}$ is a complete orthonormal system in $L^2(S^1, d\theta)$. Denote it by $\{\varphi_n(\theta), n \geq 0\}$. The Lévy Brownian motion $B(a), a \in R^2$ can be written as $X(r,\theta), a = (r,\theta)$.

Set

$$X_n(t) = \int_0^{2\pi} X(t,\theta)\varphi_n(\theta)d\theta, \; n \geq 0. \tag{3.6.1}$$

We can see that $\{X_n(t), n \geq 0\}$ is an independent system of Gaussian processes since the covariance

$$E(X_m(t)X_n(s)) = 0, \quad m \neq n,$$

for any s and t.

Let the canonical representation of $X_n(t)$ for two-dimensional parameter case be

$$X_n(t) = \int_0^t F_n(t,u)\dot{B}(u)du.$$

By using McKean's expansion[116], the canonical kernels $F_n(t,u)$'s of the representations can be obtained as

$$F_n(t,u) = \frac{1}{2\sqrt{\pi}}\left(\frac{u}{t}\right)^{l-1}\left(1 - \frac{u^2}{t^2}\right)^{\frac{1}{2}}, \quad l > 0, \quad n = 2l-1.$$

Further set

$$F_n(t,f) = \int_0^t F_n(t,u)f(u)du, \quad f \in C[0,\infty).$$

Note that

$$t^l F_n(t,f) = \frac{1}{2}\int_0^{t^2} v^{\frac{l}{2}-1}\sqrt{t^2 - v}\, f(\sqrt{v})dv,$$

and then we can prove the formula

$$\Gamma(\frac{3}{2})(D_{t^2})^{\frac{3}{2}} t^l F_n(t,f) = \frac{1}{2}t^{l-2}f(t).$$

Thus we have

$$L_t^{(n)} F_n(t,f) = f(t),$$

and so

$$L_t^{(n)} X_n(t) = \dot{B}_n(t), \tag{3.6.2}$$

where

$$L_t^{(l)} = \sqrt{\pi}\, t^{-l+2}\, D_{t^2}^{\frac{3}{2}}\, t^l, \quad D_t = \frac{d}{dt}. \tag{3.6.3}$$

This means that the white noise $\dot{B}_n(t)$ can be formed from $X_n(t)$, with $n = 2l-1, \ l > 0$.

It can easily be seen that $\{\dot{B}_n(t)\}$ is an independent system of white noises since $\{X_n(t)\}$ is an independent system.

It should be noted that for even dimensional parameter case, the innovation is obtained by applying the operator $L_t^{(l)}$, to Lévy's Brownian motion, which is not local, but causal (i.e. $\dot{B}_n(t)$ is formed from the values of $X_n(s), s \leq t$).

We now consider for the case $d = 3$. The kernel function of the representation of $X_n(t)$ is known as

$$F_n(t,u) = \begin{cases} 1 - \frac{u}{t}, & l = 0, \\ \frac{1}{\sqrt{\pi}} \left(\frac{u}{t}\right)^{l-1} \left((1 - \frac{u^2}{t^2}\right)^{\frac{1}{2}}, & l > 0, \end{cases}$$

for $n = 1$ if $l = 0$ and $n = 2l - 1$ if $l > 0$.

It is easy to show that there exists a local operator

$$L_t^{(l)} = \frac{1}{t^{l-1}} \frac{d}{dt} \frac{1}{2t} \frac{d}{dt} t^{l+1}, \tag{3.6.4}$$

such that

$$L_t^{(l)} X_n(t) = \dot{B}_n(t).$$

See the literature[140].

Remark 3.7 *For odd dimensional parameter case, say R^{2p+1}-parameter case, $M(t)$ process is $(p+1)$-ple Markov process which is known (see e.g. Lévy[105]).*

3.7 Gaussian random fields

To fix the idea, we consider a Gaussian random field $X(C)$ parametrized by a smooth convex contour in R^2 that runs through a certain class $\mathbf{C}$. It is topologized by the usual method using the Euclidean metric. Denote by $W(u), u \in R^2$, a two-dimensional parameter white noise. Let (C) denote the domain enclosed by the contour C.

Given a Gaussian random field $X(C)$ and assume that it is expressed as a stochastic integral of the form:

$$X(C) = \int_{(C)} F(C,u)W(u)du, \tag{3.7.1}$$

where $F(C,u)$ be a kernel function which is locally square integrable in u. For convenience we assume that $F(C,u)$ is smooth in (C,u). The integral is a *causal* representation of the $X(C)$.

The canonical property for a Gaussian random field can be defined as a generalization of a representation for a Gaussian process. The basic idea of the definition is the use of conditional expectation.

The stochastic variational equation for this $X(C)$ is of the form

$$\delta X(C) = \int_C F(C,s)\delta n(s)W(s)ds + \int_{(C)} \delta F(C,u)W(u)du. \tag{3.7.2}$$

In a similar manner to the case of a process $X(t)$, but somewhat complicated manner (either using the system $\{\delta n(s)\}$ describing δC or the infinite dimensional rotation group $O(E)$ that will be discussed in Chapter 5), we can form the innovation $\{W(s), s \in C\}$. For detail see the monograph[71].

Example 3.10 A variational equation of Langevin type.

Given a stochastic variational equation

$$\delta X(C) = -X(C)\int_C k\delta n(s)ds + X_0\int_C v(s)\partial_s^*\delta n(s)ds, C \in \mathbf{C},$$

where $\mathbf{C}$ is taken to be a class of concentric circles, v is a given continuous function and ∂_s^* is the adjoint operator of the differential operator ∂_s.

A solution is given by:

$$X(C) = X_0\int_{(C)} \exp[-k\rho(C,u)]\partial_u^* v(u)du,$$

where ρ is the Euclidean distance.

Remark 3.8 *Similar discussion can be given for* $C \in \mathbf{C}$, *where* $\mathbf{C} = \{\text{smooth ovaloids in } R^d, d > 2\}$. *We would like to note the significance of the roles of innovation for Gaussian random fields.*

Chapter 4

Linear processes and linear fields

The content of this chapter is concerned with linear processes as a generalization of Gaussian process and Poisson process and is based mainly on P. Lévy's ground breaking works on linear processes done around '50s. One may wonder why his classical works are still referred at present stage. In fact, he is a pioneer in this field, and his original papers are still relevant even today. One of the typical examples concerning a linear process can be seen in the literature[106].

While we were working on white noise theory, we have often met opportunities to think of foundation of stochastic process (in particular, that of Gaussian processes). Those considerations have invited us to a profound study on how to define a stochastic process. Particular interest occurred when we studied a stochastic process with linear structure, where the *dependence* of randomness is characteristic although it is simpler.

Then, we came to investigations of the linear actions or operations on those stochastic processes. So far as dependence is concerned, linearly dependent property should come right after the consideration on Gaussian case for which linearity plays a dominant role to study the analysis.

The first attempt was, therefore done for the linear operations acting on elemental processes. We then proceeded to the stage of being in search of a class of processes that can be characterized by linear operations. It is shown that the class that should be sitting next to Gaussian system has to be the class of *linear processes*. This chapter would therefore serve to see how the theory of linear processes has developed in line with white noise theory as a next class to Gaussian processes.

4.1 Gaussian systems

First we recall the definition of a Gaussian system, which was given in Chapter 3. A system of real random variables $\boldsymbol{X}= \{X_\alpha(\omega)\}$ is called a *Gaussian system*, if it satisfies the condition that any finite linear combination of members in the system is Gaussian. There *linear* combination should be emphasized.

For a Gaussian system $\boldsymbol{X}$, each member X_α is a vector valued Gaussian random variable in $L^2(\Omega, P)$. The closed subspace spanned by X_α's again forms a Gaussian system.

Before stating a basic theorem, we prove the following lemma due to P. Lévy.

Lemma 4.1 *(P. Lévy[106]) Let $f_j, j = 1,2$, be non-constant, continuous functions, and let a, b, c, d, be non-zero constants such that $ad - bc \neq 0$. If there exist functions $g_j, j = 1,2$, such that the following functional equation*

$$f_1(ax+by)f_2(cx+dy) = g_1(x)g_2(y), \tag{4.1.1}$$

holds, then, $\log|f_j|$ and $\log|g_j|, j = 1,2$, are all polynomials of degree at most 2.

Proof. Obviously $g_j, j = 1,2$, are continuous. The variables $ax + by$ and $cx + dy$ are viewed as independent variables by the assumption $ad - bc \neq 0$ and $f_j, g_j, j = 1,2$, never vanish. In particular, we can therefore assume that all the functions are positive. Set

$$\log f_j = v_j, \quad \log g_j = w_j, \quad j = 1,2.$$

Then, $v_j, w_j, j = 1,2$, are continuous and satisfy the equation

$$v_1(ax+by) + v_2(cx+dy) = w_1(x) + w_2(y).$$

Since all those functions are locally integrable, we may apply the regularization in the sense of generalized functions, for instance,

$$\int (v_1(ax+by) + v_2(cx+dy))\,\xi(y)dy = w_1(x)\int \xi(y)dy + \int w_2(y)\xi(y)dy,$$

where ξ is in the Schwartz space S.

Hence, by taking such ξ as $\int \xi(y)dy \neq 0$, w_1 is proved to be a $\mathbf{C}^\infty$-function, which comes from regularization by convolution. Similarly, we prove that w_2 is a $\mathbf{C}^\infty$-function.

Noting that $ax+by$ and $cx+dy$ are viewed as two independent variables, let them denote by s and t, we are given an equation of the same type as (4.1.1). Hence, it is proved that $f_j, j = 1, 2$, are $\mathbf{C}^\infty$-functions.

By applying $\frac{\partial^2}{\partial x \partial y}$ to the equation (4.1.1), we have

$$abv_1''(ax + by) + cdv_2''(cx + dy) = 0.$$

This means that $abv_1''(s) + cdv_2''(t) = 0$ for independent variables s and t. Hence, v_1 and v_2 should be constants, which proves the assertion.

■

By using the above lemma we can prove a remarkable theorem that characterizes a Gaussian system.

Theorem 4.1 *(P.Lévy[106]). Let X and Y be two random variables such that there exist random variables U and V such that U is independent of X and V is independent of Y, satisfying*

$$Y = aX + U \tag{4.1.2}$$

$$X = bY + V, \tag{4.1.3}$$

for constants a and b. Then, one of the following three cases holds:

(i) (X, Y) is a Gaussian system,
(ii) X and Y are independent,
(iii) there exists an affine relation between X and Y.

Proof. Consider the characteristic function $\varphi(t_1, t_2)$ which is given by

$$\varphi(t_1, t_2) = E(e^{i(t_1 X + t_2 Y)}).$$

By using (4.1.2) and (4.1.3), we have

$$E(e^{i(t_1 + at_2)X + it_2 U}) = E(e^{i(bt_1 + t_2)Y + it_1 V}),$$

and it implies

$$E(e^{i(t_1+at_2)X})E(e^{it_2U}) = E(e^{i(bt_1+t_2)Y})E(e^{it_1V}).$$

This can be expressed in the form

$$\varphi_X(t_1 + at_2)\varphi_U(t_2) = \varphi_Y(bt_1 + t_2)\varphi_V(t_1),$$

where $\varphi_x, \varphi_U, \varphi_Y$ and φ_V are characteristic functions of X, U, Y and V, respectively. This is a similar formula to (4.1.1), if we exclude a special case where $ab = 1$, that is $X = Y$. Thus, we can come to the conclusion of the lemma. Thus, the theorem is proved.

■

It should be noted that the following well-known theorem (called Lévy-Cramér Theorem, see Hida[40] Section 1.7) is quite fundamental in the study of Gaussian system.

Theorem 4.2 *Let two independent random variables X and Y be given. If the sum $X+Y$ is Gaussian, then each random variable is Gaussian except trivial cases.*

Remark 4.1 *One may expect that similar properties to those stated in Theorems 4.1 and 4.2 could hold for the Poisson case. As for Theorem 4.1, there is a counter part. The Raikov Theorem (1938) says that if the sum of two independent random variables has Poisson distribution, then each random variable is subject to Poisson distribution (one of them may be a constant). While, assertion like Theorem 4.1 does not hold for Poisson case, as is illustrated by the following example.*

Example 4.1 Let $P(t), t \geq 0$, be a Poisson process. Take $P(s)$ and $P(t)$ with $s < t$. Then

$$P(t) = P(s) + (P(t) - P(s))$$

is fine, where $P(s)$ and $(P(t) - P(s))$ are independent. On the other hand, if we have

$$P(s) = \mu P(t) + V,$$

then by the Raikov Theorem, both $\mu P(t)$ and V must be Poisson random variables. Hence, $\mu = 1$, which is not acceptable.

It seems fitting for the study of general Gaussian system to write explicit form of probability density function $p(x), x = (x_1, x_2, \cdots, x_n)$ of n-dimensional Gaussian random vector $\boldsymbol{X} = (X_1, X_2, \cdots, X_n)$.

Let $m = (m_1, m_2, \cdots, m_n)$ be the mean vector; $E(X_j) = m_j, 1 \le j \le n$, and let $V = (V_{j,k})$ be the covariance matrix; $E((X_j - m_j)(X_k - m_k))$, which is symmetric and positive definite. To fix the idea we assume that $\boldsymbol{X}$ is non-degenerated, that is the covariance matrix V is non-singular. With these notations and assumption, $p(x)$ is given by

$$p(x) = \frac{1}{\sqrt{(2\pi)^n |V|}} \exp[-\frac{1}{2}(x-m)V^{-1}(x-m)'],$$

where $x \in R^n$ and $|V|$ is the determinant of V.

Note that the Gaussian distribution is completely determined by mean vector and covariance matrix.

Example 4.2 Two dimensional Gaussian distribution.

The density function is of the form:

$$p(x,y) = \frac{(1-\gamma^2)^{-\frac{1}{2}}}{2\pi\sigma_1\sigma_2} e^{\frac{1}{2(1-\gamma^2)}\left(\frac{(x-m_1)^2}{\sigma_1^2} - 2\gamma\frac{(x-m_1)(y-m_2)}{\sigma_1\sigma_2} + \frac{(y-m_2)^2}{\sigma_2^2}\right)}, \quad (4.1.4)$$

where γ is the correlation coefficient.

As a consequence of this formula, we see that for a Gaussian system $\{\boldsymbol{X}, \boldsymbol{Y}\}$, the orthogonality of $X - E(X)$ and $Y - E(Y)$ in $L^2(\Omega, P)$ implies the independence of X and Y.

Proposition 4.1 *Let X and Y be independent random variables such that $E(X) = E(Y) = 0$. If the joint distribution of (X, Y) is rotation invariant, then $\{X, Y\}$ is Gaussian. ($X = Y = 0$ is excluded.)*

Proof. Let $\varphi(z), \varphi_1(z)$ and $\varphi_2(z), z \in R$, be characteristic functions of (X, Y), X and Y, respectively. Then $\varphi(z_1, z_2)(= E(e^{i(z_1X_1 + z_2X_2)}))$ is continuous and rotation invariant, so that it is a function of $z_1^2 + z_2^2$. Hence

$$\varphi(z_1, z_2) = \tilde{\varphi}(z_1^2 + z_2^2) = \varphi_1(z)\varphi_2(z).$$

Using $\varphi_1(0) = \varphi_2(0) = 1$, we may write

$$\tilde{\varphi}(z_1^2 + z_2^2) = \tilde{\varphi}_1(z_1^2)\tilde{\varphi}_2(z_2^2),$$

where $\tilde{\varphi}_i(z_i^2) = \varphi_i(z_i), i = 1, 2$.

Hence, φ, φ_1 and φ_2 are exponential functions of z^2. They must be characteristic functions of Gaussian distributions. ■

A Gaussian system, which we are now going to discuss, may involve many random variables and be dealt with as a vector space over R, by replacing orthogonality with independence. It satisfies the following conditions:

1) Each member is Gaussian in distribution (by definition), and
2) for any pair of members, linear relation always hold, in the manner shown in the above Theorem 4.1.

The second property can be generalized later as in Theorem 4.3.

Definition 4.1 Let $\boldsymbol{X}= \{X_\alpha\}$ be a system of random variables X_α. A random variable Y that is outside of $\boldsymbol{X}$ is said to be *linearly dependent* on $\boldsymbol{X}$ if Y is expressed in the form

$$Y = U + V,$$

where U is a member in $\boldsymbol{X}$, and where V is independent of $\boldsymbol{X}$.

In the above expression U and V are unique up to non-random constants.

Note that the (continuous) linear function is formed in the topology of $L^2(\Omega, P)$ and that both U and V are uniquely determined up to an additive constant.

Definition 4.2 A system $\boldsymbol{X}$ is said to be *completely linearly correlated*, if, for any subclass $\boldsymbol{X}' \subset \boldsymbol{X}$, any element Y which is outside of $\boldsymbol{X}'$ is linearly dependent on $\boldsymbol{X}'$.

Theorem 4.3 *A completely linearly correlated system $\boldsymbol{X}$ is Gaussian except for trivial subclass which involves only mutually independent members.*

The above Theorem 4.3 can be stated in a more precise manner.

Theorem 4.4 *Let a collection* $\boldsymbol{X} = \{X_\alpha, \alpha \in A\}$ *be a completely linearly correlated system with* $E(X_\alpha) = 0$ *for every* α. *Then, the parameter set* A *is decomposed into disjoint sets:*

$$A = A_0 + \sum_{n \geq 1} A_n,$$

where
1) $\mathbf{X}_0 = \{X_\alpha, \alpha \in A_0\}$ *is a Gaussian system,*
2) the systems $\mathbf{X}_n = \{X_\alpha, \alpha \in A_n\}, n \geq 0$, *are mutually independent,*
3) for $n > 0$, $\mathbf{X}_n$ *involves only one element, except constant.*

Proof. We may exclude those X_α's such that $P(X_\alpha = 0) = 1$. Suppose there exists at least one Gaussian random variable, say X_{α_0}. There exists a maximal subset A_0 of the $S(\subset A)$ such that $\{X_\alpha, \alpha \in S\}$ is Gaussian. Obviously $\{X_\alpha, \alpha \in A_0\}$ which involves X_{α_0}, is Gaussian.

Now take $\alpha \in A_0$ and $\beta \in A_0^c$ arbitrarily. By the definition of A_0, the system $\{X_\beta, X_\alpha, \alpha \in A_0\}$ is definitely not Gaussian. By the assumption that the system $\boldsymbol{X}$ is a completely linearly correlated, it is possible to have a decomposition

$$X_\alpha = aX_\beta + V,$$

where X_β and V are independent, and $V \neq 0$, $a \neq 0$. Then, the right-hand side is a sum of independent random variables and is Gaussian in distribution. Use Theorem 4.2 to show that this fact contradicts to the assumption that X_β is not Gaussian. Hence, a must be 0. Further it can be proved that X_β is independent of the system $\{X_\alpha, \alpha \in A_0\}$.

Now suppose there are two members X_β and $X_{\beta'}$ such that $\beta, \beta' \in A_0^c$. They cannot be Gaussian, but linearly correlated. So, Theorem 4.1 says they are in the exceptional case where either in a linear relation or independent.

As a result, $\boldsymbol{X}$is a direct sum of the systems $\{X_\alpha, \alpha \in A_j\}$, $j \geq 0$. Thus the theorem has been proved.

■

Note. The members in $\boldsymbol{X}$ are not necessarily linearly ordered.

It seems to be a good time to recall the canonical representation theory of a Gaussian process $X(t), t \geq 0$. Let it be expressed in the form which is a canonical representation:

$$X(t) = \int_0^t F(t,u)\dot{B}(u)du,$$

based on a white noise $\dot{B}(t)$, although each $\dot{B}(t)$ is not an ordinary random variable. However, we are suggesting to extend Theorem 4.4 to a suitable system of generalized random variables. The $\dot{B}(t)$ eventually becomes the *innovation* of the $X(t)$. See Chapter 8.

With this consideration we can extend the theory of canonical representations to some extent.

A Gaussian process is said to have no *remote future.* If we set $\mathbf{B}^t(X) = \mathbf{B}(X(s), s \geq t)$, then the remote future is represented by the sigma-field $\mathbf{B}^\infty(X) = \bigwedge_t \mathbf{B}^t(X)$. So, no remote future means that $\mathbf{B}^\infty(X)$ is a trivial field, i.e. equal to the set $\{\Omega, \emptyset\}$ (mod 0). In such a case, we may consider a *backward canonical representation* of the form

$$X(t) = \int_t^\infty F^-(t,u)\dot{B}^-(u)du,$$

where $\dot{B}^-$ is a white noise such that the equality $\mathbf{B}^t(\dot{B}^-) = \mathbf{B}^t(X)$ holds for every t.

We can state

Proposition 4.2 *If a Gaussian process $X(t)$ is stationary and mean continuous, and if it has no remote future, then there exists the unique backward canonical representation based on a white noise $\dot{B}^-(t)$ with an expression*

$$X(t) = \int_t^\infty F(u-t)\dot{B}^-(u)du.$$

The white noise $\dot{B}^-(t)$ appeared in the above formula is called *the backward innovation.*

Interesting properties of the backward canonical representation can be obtained as counterparts of the canonical representation established in Chapter 3.

Let us now return to the (forward, i.e. ordinary) canonical representation theory to state some supplement to the last chapter.

One of the significant properties of the canonical representation theory can be seen in the application to the study of way of dependence of a stochastic process, in particular multiple Markov property, say N-ple Markov property. We have seen that the property can be characterized in terms of the analytic properties of the canonical kernel (see Chapter 3 and the literature[35]). Those properties can be applied to the prediction theory as well as filtering of signal in electrical engineering.

One can also speak of an information theoretical properties of such a process. To fix the idea, the N-ple Markov process is assumed to be stationary. Then, it is proved (see Win Win Htay[167]) that, as for the amount of information carried by the process, the higher the multiple Markov property, the more the information is lost. This fact is one of the interpretations on how the order of multiple Markov property of a Gaussian process describes the degree of dependence.

To give a concrete example, a stationary multiple Markov process was taken with the canonical representation of the form

$$X(t) = \int_{-\infty}^{t} \sum_{1}^{N} a_k \exp[-k\alpha(t-u)]\dot{B}(u)du, \quad \alpha > 0,$$

which is most typical from many viewpoints. (See Win Win Htay[167].) As an additional note, one may say that this is a stationary version of the multiple Markov Gaussian process $M(t)$ which is obtained by taking the average of the Lévy's Brownian motion over the sphere of radius t.

4.2 Poisson systems

We now come to a Poisson process $P(t), t \geq 0$, which is a typical example of an additive process that immediately comes after Brownian motion. Poisson process has lots of similar probabilistic properties to Brownian motion. In some sense, Brownian motion and Poisson process are in a dual position. However we see many dissimilarity, as well. One of the crucial differences between them can be seen in the following example from the viewpoint of the completely linearly correlated property.

Example 4.3 Let $P(t)$ be a Poisson process and set $\boldsymbol{X}= \{P(t), t \geq 0\}$. Take $\boldsymbol{X}'$ to be $\{P(v), v \leq s\}$ and take $Y = P(t)$ for $t \geq s$. We take $X = P(s)$ and $U = P(t) - P(s)$ to see if the completely linearly correlated property holds or not,

On the other hand, if we take $\boldsymbol{X}''= \{P(v), v \geq t\}$ instead of $\boldsymbol{X}'$ and if $Y = P(s)$ with $s \leq t$, then a decomposition like

$$P(s) = \int_t^\infty f(u)\dot{P}(u)du + V$$

is impossible. For, noting that $E(V^2) < \infty$, we can see that the correlation between $\int f(u)\dot{P}(u)du$ and $P(s)$ violates the equality.

Another way to observe the linearly correlated property is as follows: Gaussian system has a linear structure, as it were, in both directions, towards the future and to the past. For the case of a Gaussian process, we have forward and backward canonical representations.

On the other hand, Poisson process has the linear structure only in one direction as is shown in the last example. We may say that it is one-sided *linearly correlated.*

Keeping what are discussed regarding linearly dependence in mind, we wish to generalize the linear property to somewhat wider class of stochastic processes. This is one of the motivations of introducing the class of linear processes later in Section 4.4.

4.3 Linear functionals of Poisson noise

Recall that a Poisson noise, denoted by $\dot{P}(t), t \in [0, \infty)$, is a generalized stochastic process, the characteristic functional of which is

$$C_P(\xi) = \exp[\lambda \int_0^\infty (e^{i\xi(u)} - 1)du],$$

where λ (> 0) is the *intensity.*

Like a relationship between white noise and Brownian motion, we can form a stochastic process that is expressed as an integral of a Poisson noise

$\dot{P}(t)$. More generally, let $X(t)$ be given by

$$X(t) = \int_0^t G(t,u)\dot{P}(u)du, \tag{4.3.1}$$

where the kernel $G(t,u)$ is assumed to be smooth in the variable (t,u). We consider such a process by analogy with a Gaussian process represented by a white noise.

For a moment, the infinitesimal random variable $\dot{P}(u)du$ is viewed as an orthogonal random measure, so that we may write it as $dP(u)$. Another way of understanding $\dot{P}(t)$ is that the smeared variable $\dot{P}(\xi), \xi \in E$, is a generalized stochastic process with independent value at every t, in the sense of Gel'fand. (Compare with the interpretation of $\dot{B}(t)$.) It is noted that we often let $\dot{P}(t)$ be centered, i.e. it is replaced by $\dot{P}(t) - \lambda$, where λ is the intensity (expectation). Although there are many ways of defining the integral (4.3.1), we always have the same probability distribution regardless of the expression.

The notion of the canonical representation and the canonical kernel are defined in the same manner to the Gaussian case. If we use the same notation for the sigma-fields, canonical property can be defined as in Section 3.2. The equality

$$\mathbf{B}_t(X) = \mathbf{B}_t(\dot{P}) \tag{4.3.2}$$

for every t implies the canonical representation. It is noted that they are defined in the sense of linear analysis. Important remark is that the kernel criterion for the canonical kernel is different, as is seen in the next example.

Example 4.4 Suppose that for the representation (4.3.1) $G(t,t) \neq 0$ for every t, then we can find all the jump points of *sample functions* of $P(t)$. Hence we can prove the equality (4.3.2) of the sigma-fields.

Note that the kernel is smooth in u we may define the above integral as a sample function-wise (of $P(u)$) Stieltjes integral.

4.4 Linear processes

Following the idea of P. Lévy[106] the definition of a linear process is now given as follows.

Definition 4.3 A stochastic process $X(t), t \in [0, \infty)$, is called a *linear process*, if it is expressed as a sum

$$X(t) = X_G(t) + X_P(t), \tag{4.4.1}$$

where

$$X_G(t) = \int_0^t F(t,u)\dot{B}(u)du, \quad X_P(t) = \int_0^t G(t,u)\dot{Y}(u)du. \tag{4.4.2}$$

In the above expression $B(u)$ is a Brownian motion and $Y(u)$ is a compound Poisson process (see Hida[36] Section 3.2) with stationary independent increments. Two processes are assumed to be independent. The integral giving $X_P(t)$ is defined by the sample function-wise integral, where the topology to define the integral is taken to be convergence in probability.

A linear process satisfies the property of linearly correlated and is viewed as a generalization of a Gaussian process from the viewpoint of linearly correlated property.

Theorem 4.5 *Let $X(t)$ be given by the above integrals. Assume that the kernels $F(t,u)$ and $G(t,u)$ are canonical kernels (in the sense of representation of Gaussian processes) with smooth $G(t,t)$ such that $G(t,t) \neq 0$. Then, the sample functions of the $X(t)$ can determine the two components $X_G(t)$ and $X_P(t)$.*

Proof. Take a characteristic functional of $X(t)$, which can be factorized by those of $X_G(t)$ and $X_P(t)$. Then, apply the functional derivative and use analytic methods, we can compute the kernels $F(t,u)$ and $G(t,u)$. Note that the canonical property of the kernels plays an important role. The rest of the proof is rather obvious.

■

For further details on this topic, see the joint work with Win Win Htay[153], where more general theory is developed in details.

There is another approach to a linear process. A certain linear process may be thought of as a mathematical model of a system in communication theory.

Assume that the input to a communication system consists of Gaussian white noise and Poisson noise with intensity λ, which are denoted by $\dot{B}(t)$

and $\dot{P}(t)$, respectively. Further assume that the input passes through an unknown but bounded continuous filters $F(t,u)$ and $G(t,u)$, respectively. The filters are Volterra type so they should be causal and are assumed to be causally invertible, i.e. canonical. The mathematical expression of the output $X(t), t \geq 0$ is the same as $X_G(t) + X_P(t)$ given by (4.4.1), where $Y(t)$ is simply a standard Poisson process with unit jump. We observe a sample path on infinite time interval. Then, we can appeal to the ergodic theorem and are able to compute the characteristic functional $C(\xi)$ as the time average. It is of the form

$$C(\xi) = \exp\left[-\frac{1}{2}\|F * \xi\|^2 + \lambda \int (e^{i(G*\xi)(u)} - 1)du\right], \tag{4.4.3}$$

where $(F*\xi)(u)$ is defined by $\int F(t,u)\xi(u)du$, similarly for $(G*\xi)(u)$. The $C(\xi)$ never vanishes, so that we can determine $c(\xi) = \log C(\xi)$ uniquely with $c(0) = 0$

$$C(\xi) = -\frac{1}{2}|F * \xi|^2 + \lambda \int (e^{i(G*\xi)(u)} - 1)du. \tag{4.4.4}$$

Take the variation $\delta C(\xi)$ which is of the form

$$\delta C(\xi) = -\int (F*\xi)(u)du \int F(t,u)\delta\xi(u)du + \lambda i \int e^{i(G*\xi)(u)}(G*\delta\xi)(u)du.$$

Set $\xi = 0$, and let $\delta\xi$ vary. Then we can determine the kernel $\lambda G(t,u)$. Coming back to (4.4.4) we can determine $\int \left(\int F(t,u)\xi(t)dt\right)^2 du$, the second variation of which determines the kernel $F(t,u)$ for every u.

From $\|F * \xi\|^2 = \int \left(\int F(t,u)\xi(t)dt\right)^2 du$, we know

$$\int \left(\int F(t,u)\xi(t)dt\right)\left(\int F(s,u)\eta(s)ds\right)du = \int\int \Gamma(t,s)\xi(t)\eta(s)dtds,$$

where

$$\Gamma(t,s) = \int F(t,u)F(s,u)du.$$

Since ξ and η are arbitrary we can determine $\Gamma(t,s)$, which is the covariance function of $X_G(t)$. The factorization of $\Gamma(t,s)$ gives us the canonical kernel $F(t,u)$ up to equivalence.

Summing up what have been obtained, we have a characteristic functional version of the results of Theorem 4.5.

Theorem 4.6 *Given a characteristic functional $C(\xi)$ by the formula (4.4.3), we can then determine canonical kernels $F(t,u)$ and $G(t,u)$. Further we can form a version of a linear process expressed by the formulas (4.4.1) and (4.4.2) on a measure space $(E_G^* \times E_P^*, d\mu \times d\mu_P)$, where $d\mu$ and $d\mu_P$ are white noise measure on E_G^* and Poisson noise measure on E_P^*, respectively.*

Example 4.5 Let $X(t), t \geq 0$ be a sum of Brownian motion $B(t)$ and Poisson process $P(t)$:

$$X(t) = B(t) + P(t).$$

Then, Theorem 4.5 is applied. As for the sigma-fields, we have

$$\mathbf{B}_t = \mathbf{B}_t(B) + \mathbf{B}_t(P).$$

Theorem 4.6 can be extended to the case of a generalized stochastic process.

Example 4.6 $X(t) = \dot{B}(t) + \dot{P}(t)$, we use characteristic functional rather than sample path. Discrimination of two components is obvious in terms of characteristic functional.

Finally, in this chapter, it seems better to give a short remark on the higher dimensional parameter case. A Gaussian field with higher dimensional parameter, e.g. the R^d-parameter Lévy Brownian motion, requires additional considerations. For instance, multiplicity problem, which will occur when we consider the study of representation in terms of white noise, involves in more complex manner. It is recommended to take the radial direction in R^d to be the main direction of propagation, and rotations would serve to provide multiplicity and to give information in different directions.

For a higher dimensional parameter process, that is a random field, of Poisson type, representation in terms of Poisson noise seems easier to be investigated in order to have linear field of Poisson type.

Thus, one can see, intuitively, the complexity of a random field with higher dimensional parameter, when we wish to generalize the linear process technique (see Chapter 7).

4.5 Lévy field and generalized Lévy field

Lévy field is a random field that generates various non-Gaussian elemental stochastic processes of Poisson type. Its characteristic functional $C_L(\xi)$ would be of the following form

$$C_L(\xi) = \exp[\int_0^\infty dt \int_{R-\{0\}} du \ (e^{i\xi(t,u)} - 1)],$$

where $\xi(t,u)$ is a test function defined on $E = K([0,\infty)) \times K(R - \{0\})$, K being the test function space in the sense of Gel'fand-Vilenkin[30]. It looks like an R^2-parameter Poisson noise, but not quite. Anyhow, $C_L(\xi)$ is a characteristic functional, so that there exists a generalized random field (a two-dimensional parameter random field) $X(t,u)$, which determines the probability measure ν on E^*. A sample function $X(t,u)$ is denoted by $x(t,u)$.

The first question on $X(t,u)$ is how it generates a generalized, in fact elemental, stochastic process. To answer this question, we consider a functional of sample function of $X(t,u)$ expressed in the form

$$y(t,u) = f(u) \int x(t,v)\delta_1(v)dv,$$

where $f(u)$ is continuous and locally bounded on $R-\{0\}$. There is a stochastic process with such sample functions $y(t,u)$, u being fixed. In particular, if $f(u) = u$, then we are given a Poisson noise multiplied by u, namely $u\dot{P}(t)$. Hence, we can say that all the elemental Poisson noises can be obtained from $X(t,u)$.

We claim that the $X(t,u)$ plays the role of a *generator* of elemental Poisson noises of various magnitudes of jumps. Our basic step "reduction" is applied to the Lévy field $X(t,u)$ of Poisson type to guarantee that the field in question possesses this property.

Next question is jump finding. Observing a sample function of $X(t,u)$, one should think of some technique to determine $y(t,u)$. Intuitive treatment may not guarantee the measurability of the trick.

Then, we may find out how to compute a marginal distributions of ν. It can be obtained by restricting the parameter of the variable of the characteristic functional $C_L(\xi)$. The answer is not simple. We would like to refer to our paper[70] for details.

4.6 Gaussian elemental noises

We can form a compound Gaussian noise consisting of independent Gaussian noises with various different variances. Assume that all the components are centered. Namely, a *compound Gaussian noise* is a superposition of various Gaussian noises with mean zero. It can be determined by the characteristic functional which is expressed in the following form

$$C_G(\xi) = \int_0^\infty \exp\left[-\frac{\sigma^2}{2}\|\xi\|^2\right] dm(\sigma),$$

where m is probability measure that is supported by the positive half line with $m(\{0\}) = 0$. The uniquely determined probability measure on E^* by $C_G(\xi)$ is denoted by μ_G.

Note. Further understanding of the meaning of the above decomposition will be given in the next chapter.

We can introduce a compound Gaussian noise which has the characteristic functional $C_G(\xi)$.

This characteristic functional defines a compound Gaussian measures μ_σ in the sense that

$$\mu_G(A) = \int \mu_\sigma(A) dm(\sigma),$$

where each μ_σ is elemental in the sense that it is atomic. Indeed, it is *ergodic.*

Here we give a short note concerning the decomposition of the μ into μ_σ's. Basically the strong law of large numbers is used. Take a complete orthonormal system $\{\xi_n\}$ in $L^2(R^n)$.

The measurable set

$$A_\sigma = \left\{x : \lim_{N=\infty} \sum_1^N \langle x, \xi_n\rangle^2 = \sigma^2\right\}$$

has μ_σ-measure 1, i.e. $\mu_\sigma(A_\sigma) = 1$, by the strong law of large numbers. Further rigorous interpretation is omitted.

Chapter 5

Harmonic analysis arising from infinite dimensional rotation group

This chapter is devoted to the harmonic analysis that arises from the infinite dimensional rotation group, which is one of our favorite and, in fact, significant characteristic tools used in the white noise analysis. The harmonic analysis presents one of the main aspects of white noise theory. Recall that the other characteristic is the space of generalized white noise functionals that have been discussed in Chapter 2. One more characteristic, if permitted, is seen in the effective use of white noise to discuss innovation of stochastic processes as well as random fields in the same manner.

The readers will see, in what follows, the privilege to have useful tools of the harmonic analysis that is established by using the infinite dimensional rotation group.

5.1 Introduction

One may wonder why rotation group is involved in white noise analysis. To answer this question, we can state various plausible reasons, which are now in order.

i) First we appeal to the strong law of large numbers. We have seen in Chapter 2, say Y_n's in Section 2.4, also we will see in Section 5.3, there exist countably many number of independent random variables defined on white noise measure space that are subject to the standard Gaussian distribution. The sigma-field, with respect to which all those random variables are measurable, is in agreement with the sigma-field of white noise measure space. At the same time those random variables are viewed as coordinates. It can therefore be shown, intuitively speaking, that the white noise measure μ is a uniform measure on an infinite dimensional sphere with radius $\sqrt{\infty}$, if we speak intuitively.

ii) The characteristic functional $C(\xi)$ of μ is a function of the square of the L^2-norm $\|\xi\|$. This fact immediately suggests to us, although in a formal level, that the characteristic functional, and hence the measure μ, is invariant under the rotations. Thus, we establish a harmonic analysis arising from the group of those rotations.

iii) As the third reason, we may recall the interesting and surprising fact in classical functional analysis. To define a uniform measure on the unit sphere of a Hilbert space, we usually try to approximate by the uniform measure on the finite dimensional, say d-dimensional, ball V^d. The radius of the ball has to be proportional to $\sqrt{d}$. If d is getting large, the uniform measure on the sphere approximates the white noise measure μ. In fact, Hida-Nomoto[63] constructed a white noise measure by taking the projective limit of spheres and such a method of construction can be compared to that of a Poisson noise, where associated group plays an essential role in each case; the former is associated with rotation group and the latter with symmetric group.

Incidentally, such discussions lead us to infinite dimensional Laplacian in connection with rotation group.

iv) Various applications of white noise theory have stimulated the use of group to describe invariance, symmetry and other abstract properties. For instance, in quantum dynamics, group is an important tool to describe quantum properties and in the theory of information sociology they often use the rotation and symmetry groups in their study to discover latent traits.

Some other reasons may be skipped, however we claim that those facts are behind our idea of white noise analysis.

With these facts in mind, we are now going to define an *infinite dimensional rotation group* and to proceed to the analysis that can be thought of as a harmonic analysis. To illustrate the line of development of the analysis, we propose a diagram which follows right after Section 5.2. It will be a road map of the harmonic analysis that we are going to develop.

Since the infinite dimensional rotation group is quite large, indeed it is neither compact nor locally compact under reasonable topology, it seems useful to focus on significant subgroups, which are classified into two classes, namely Class I and Class II. The subgroups in Class I include rotations that are defined with the help of a complete orthonormal system in the basic Hilbert space, hence countably generated. While subgroups in Class II consists of one-parameter subgroups of diffeomorphism acting on the parameter space. Hence, those subgroups in Class II heavily depend on the geometric structures of the given parameter spaces.

As a supplementary remark on the rotation group, we assert that a rotation (in fact, its adjoint operator on E^*) acts on individual sample functions of white noise. We tacitly aim at a so-to-speak generalized harmonic analysis in the sense of N. Wiener, where the analysis is not on the space L^2, but deals with functionals $\varphi(x)$ of sample functions $x \in E^*$ directly.

Harmonic analysis related to transformation group within stochastic analysis is not necessarily limited to the case of rotation group. We would like to announce that we shall meet another interesting case where Poisson noise and symmetric group are involved being closely related and another harmonic analysis is developed. Chapter 7 is devoted to these topics. One can see both similarity and dissimilarity to a (Gaussian) white noise where some comparison of the two noises suggest to us a profound relationship from the viewpoint of harmonic analysis.

For the topics discussed in this chapter we refer the reader to the literatures[40,4,71] to see that some more details are found depending on the questions raised here, respectively.

5.2 Infinite dimensional rotation group $O(E)$

The basic nuclear space is usually taken to be either the Schwartz space or the space D_0 that is isomorphic to the space $C^\infty(S^d)$. The latter can be introduced as a subgroup isomorphic to the conformal group. Let a nuclear space E be fixed. It is assumed to be a dense subspace of $L^2(R^d)$.

Definition 5.1 A continuous linear homeomorphism g acting on E is called a *rotation* of E if the following equality holds for every $\xi \in E$;

$$\|g\xi\| = \|\xi\|.$$

Obviously, the collection of all rotations of E forms a group (algebraically) under the product

$$(g_1 g_2)\xi = g_1(g_2\xi),$$

with the identity I such that $Ig = gI = g$, for every g.

The group is denoted by $O(E)$, or by O_∞ if there is no need to specify the basic nuclear space E.

Since $O(E)$ is a group of transformations acting on the topological space E, it is quite natural to let it be topologized by the compact-open topology. Thus, we are given a topological group keeping the same notation $O(E)$ that has been introduced above.

The adjoint transformation, denoted by g^*, of $g \in O(E)$ is defined by

$$\langle x, g\xi \rangle = \langle g^* x, \xi \rangle, \quad x \in E^*, \xi \in E,$$

where $\langle \cdot, \cdot \rangle$ is the canonical bilinear form that connects E and E^*. It is easy to show that the g^* is a continuous linear transformation acting on the space E^*. The collection $O^*(E^*) = \{g^*; g \in O(E)\}$ also forms a group as is easily proved.

Proposition 5.1 *The group $O^*(E^*)$ is (algebraically) isomorphic to the rotation group $O(E)$ under the mapping*

$$g \mapsto (g^*)^{-1}, \quad g \in O(E).$$

In view of this, the group $O^*(E^*)$ can also be topologized so as to be isomorphic to $O(E)$, and the topological group $O^*(E^*)$ is also called *infinite dimensional rotation group* (if necessary, we may add, acting on E^*.

There is a fundamental theorem regarding the probabilistic role of the infinite dimensional rotation group.

Recall that the white noise measure is introduced on the measurable space $(E^*, \mathcal{B})$, where $\mathcal{B}$ is the sigma-field generated by the cylinder subsets of E^*. As before, we use the notation μ for the white noise measure determined by the characteristic functional $C(\xi) = \exp[-\frac{1}{2}\|\xi\|^2]$.

Theorem 5.1 *The white noise measure μ is invariant under the infinite dimensional rotation group $O^*(E^*)$.*

Proof. It is easy to see that g^* is a $\mathcal{B}$-measurable transformation, since it carries a cylinder subset of E^* to another cylinder subset. Hence, $gx^*\mu$ is well defined:

$$dg^*\mu(x) = d\mu(g^*x).$$

Now observe the characteristic functional of the measure $g^*\mu$:

$$\begin{aligned}\int_{E^*} \exp[i\langle x, \xi\rangle] dg^*\mu(x) &= \int_{E^*} \exp[i\langle (g^*)^{-1}y, \xi\rangle] d\mu(y) \\ &= \int_{E^*} \exp[i\langle y, g^{-1}\xi\rangle] d\mu(y) \\ &= \exp[-\frac{1}{2}\|g^{-1}\xi\|^2] = \exp[-\frac{1}{2}\|\xi\|^2],\end{aligned}$$

which is in agreement with the characteristic functional of white noise measure μ. Thus, the theorem is proved.

■

A converse of this theorem should now be discussed. We actually have the following well-known result.

Theorem 5.2 *If a probability measure ν on E^* is invariant under the group $O^*(E^*)$, then ν is a sum of delta measure $a\delta_0, a > 0$, and a sum of Gaussian measure μ_σ with variance σ^2.*

The theorem is stated more explicitly as follows

$$\nu = a\delta_0 + \int_{(0,\infty)} \mu_\sigma dm(\sigma). \tag{5.2.1}$$

For the details of the proof of the equation (5.2.1) we refer to Hida[40], but the idea is that the assumption on ν implies that the characteristic functional $C_\nu(\xi)$ is invariant under $O(E)$, so that it is a function of $\|\xi\|$ only. Also, the function is continuous and positive definite. Hence, we come to the formula

$$C_\nu(\xi) = a + \int_0^\infty \exp[-\frac{\sigma^2}{2}\|\xi\|^2] dm(\sigma), \tag{5.2.2}$$

which is nothing but the characteristic functional of the measure ν given by (5.2.1). ■

5.3 Harmonic analysis

The main topics on infinite dimensional harmonic analysis are going to be discussed in what follows in line with white noise analysis. (Sections 5.3 - 5.8). The topics are as follows:

1) Unitary representations of subgroups of $O(E)$.
2) Probabilistic roles of those subgroups.
3) Fourier transform and related topics.
4) Laplacian, actually three kinds of Laplacian.

Then follow complexification of white noise and the unitary group $U(E_c)$ in Chapter 6.

Some preliminaries and motivations should now be prepared quickly.

First let us have a naive observations of, as it were, the support of the white noise measure μ, as was mentioned in i) of Section 5.1. Take a complete orthonormal system $\{\xi_n\}$ in $L^2(R)$ to have a system of independent standard Gaussian random variables $\{\langle x, \xi_n\rangle\}$. Then, the strong law of large numbers tells us that

$$\lim \frac{1}{N}\sum_1^N \langle x, \xi_n\rangle^2 = 1, \quad a.e., (\mu),$$

which illustrates in a naive expression such that, with the coordinates $x_n = \langle x, \xi_n\rangle$,

$$\sum_1^N x_n^2 \sim N,$$

holds for μ-almost all $x = (x_n) \in E^*$.

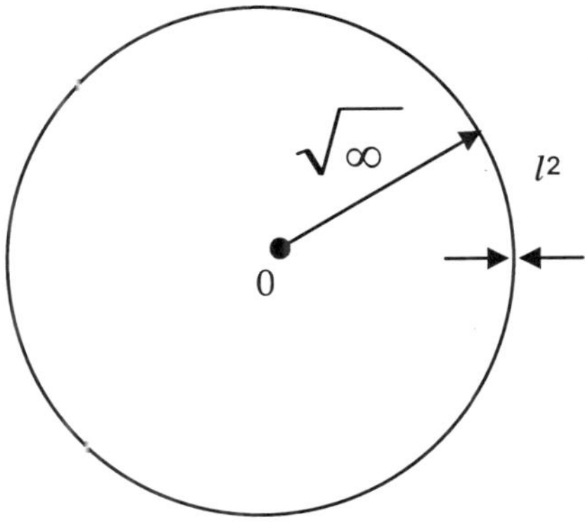

Fig. 5.1.

In other words, this approximation is true so far as x is in the support of the white noise measure μ. Hence the "support" (in an intuitive level) is approximated by $(N-1)$-dimensional sphere with radius $\sqrt{N}$. We may therefore think of the support of μ to be an infinite dimensional sphere with radius $\sqrt{\infty}$. Even more, the measure looks like a uniform (i.e. rotation invariant) measure on the sphere. Hence we can carry on the harmonic analysis where our rotation group $O^*(E^*)$ acts on the support of white noise measure μ in such a way that the μ is kept invariant.

Remark 5.1 *Coming back to the expression of a rotation invariant measure ν, it is shown that each component $\mu_\sigma, \sigma > 0$, in (5.2.1) has the support characterized by the limit*

$$\lim \frac{1}{N} \sum_1^N \langle x, \xi_n \rangle^2 = \sigma^2, \quad a.e., (\mu_\sigma).$$

Hence, we can prove that if σ and σ' are different, then μ_σ and $\mu_{\sigma'}$ are singular. The Hilbert space $L^2(E^*, \nu)$ is, as it were, a sum (in fact, integral) of the $L^2(E^*, \mu_\sigma)$.

The following fact is rather well known concerning the harmonic analysis of functions defined on the sphere S^2, and it is helpful to understand the roles of $O^*(E^*)$, where μ is viewed as the uniform measure. Consider the 3-dimensional rotation group $SO(3)$. The S^2 can be identified with the symmetric space $SO(3)/SO(2)$. The uniform measure $d\sigma$ on S^2 is defined by the surface area which is invariant under the rotations g in

$SO(3)$. What we have observed above with the help of Fig. 5.1 can be viewed as an infinite dimensional analogue of this fact.

Now further review follows. The (three dimensional) Laplace-Beltrami operator Δ_3 can be determined by the rotation group up to positive constant in such a way that

i) Δ_3 is a quadratic form of the members of the Lie algebra $SO(3)$,
ii) it commutes with $SO(n)$,
iii) it is non-negative and annihilates constants.

The eigenfunction of the Laplace-Beltrami operator span the entire Hilbert space $L^2(S^2)$, and the n-th component of the direct sum decomposition of $L^2(S^2)$ can be obtained by the eigenfunction of Δ_3 depending on n, and so forth. An infinite dimensional analogue will be established referring to Section 2.5.

Such an idea of characterizing Laplacian will be reminded in Chapter 10 Section 10, when we discuss the Volterra Laplacian Δ_V as well as the Lévy Laplacian Δ_L.

Also let us remind the Peter-Weyl theorem that guarantees a decomposition of a regular unitary representation into irreducible representations of a compact Lie group, although infinite dimensional rotation group in question is neither compact nor locally compact, but still we could see some analogy with this famous theorem in the infinite dimensional case with which we are concerned.

We know that the significant results on harmonic analysis, just reviewed above, arising from the finite dimensional rotation group would successfully be established in the case of the white noise analysis. Surely, this is true as has been expected so far, but those are certainly not covered by what are expected. Profound additional investigations are, of course, necessary, and we are even optimistic to discover important new results, noting that we are concerned with infinite dimensional case. This situation can be illustrated by the diagram shown in the following pages. There are four floors starting from Ground floor **G**. Then follow the floors I, II and III, depending on the level, needless to say that the higher, the more advanced.

Note that complexification of the rotation group is quite significant. This will be discussed in the next chapter, separately.

Ground Floor G. Finite dimensional harmonic analysis

$$L^2(S^d) = \bigoplus_n H_n,$$

$$S^d = SO(d+1)/SO(d).$$

Hyper functions on S^d, $O(n)$, $S(n)$, G_∞.

Lie group $SO(d+1)$ and spherical Laplacian Δ_d.

Peter-Weyl theorem.

Irreducible unitary representation of $SO(d)$ on H_n.

Fourier series expansion.

Floor I. Hyper-finite dimensional analysis

Fock space : $(L^2) = \bigoplus H_n$. (Wiener-Itô)

White noise functionals $\approx$ functions on $S^\infty(\sqrt{\infty})$.

Laplace-Beltrami operator Δ_∞; number operator $N = -\Delta_\infty$.

Irreducible unitary representation of $G_\infty \cong \lim SO(n)$.

Fourier-Wiener transform.

$\mathcal{T}$-transform and S-transform, integral representation.

Complex white noise, homomorphic functionals.

Diagram 1.

Floor II. Infinite dimensional analysis

Generalized white noise functionals (white noise distributions),

$$(S) \subset (L^2) \subset (S)^*.$$

$$(L^2)^+ \subset (L^2) \subset (L^2)^-.$$

Potthoff-Streit characterization of $(S)^*$

Infinite dimensional rotation group $O(E)$

G_∞, Lévy group $\mathcal{G}$, Windmill and sign-changing subgroup

Laplacian : Δ_V (or Δ_G), Δ_L

creation ∂_t^*, annihilation ∂_t and rotation; $\gamma_{s,t} = \partial_s^*\partial_t - \partial_t^*\partial_s$.

Whiskers: One-parameter subgroups that come
from diffeomorphism of parameter space.

Innovation, actions on random fields.

Tools for variational calculus.
Spectral representation, ergodic property of a one–parameter group.

Whiskers describe reversible and/or irreversible properties.
Semigroups are also introduced.

Development of path integrals: Gauge transformations.

Cooperations with noncommutative geometry.

Poisson noise, symmetric group and unitary representations.

Diagram 2.

Floor III. Ultra infinite dimensional analysis
Bridge to quantum probability.
Random field $X(C)$ living in $(S)^* \longrightarrow U(C,\xi)$.
Hadamard equation.
Quantum field — gauge field, electro-magnetic field, etc.
Tomonaga-Schwinger equation.
Quantization of fields.
Various kinds of whiskers (half whiskers).
New applications.
Applications to molecular biology.
Whiskers help to form innovation of processes and fields.
Roles of the conformal group.
Cooperation with algebraic geometry.
Complexification of rotation group and Complex white noise.
(complex Lie algebras are useful.)

Diagram 3.

There are some remarks concerned with the Diagrams.

Remark 5.2

(1) Concepts on Floor I could be approximated by members in G_∞ on the Ground floor.

(2) Floor II cannot be approximated by the finite dimensional analysis under the natural topology.

(3) Floor III includes what are essentially infinite dimensional and are not to be listed in the lower floors.

Remark 5.3 *It was 1992 when the author showed the original form of the diagram on the occasion of the seminar at University of Roma, Tor Vergata. After that the diagram has been modified according to the development of white noise theory. Thanks are due to Professor Accardi for his comments.*

Remark 5.4 *A quotation from P. Lévy (see his autobiography [109]). J'ai bôti un étage de cet édifice: que d'autres continuent !*

5.4 Addenda to the diagram

Here we wish to give various interpretations on the concepts presented in the diagram.

Note on the Diagram

First we make one important note. We claim that the group $O(E)$ should not be viewed simply as a limit of $SO(n)$ as n tends to ∞ in any sense, however we can still see some analogy to the limit of finite dimensional rotations as can easily be found out. But we should be careful to see the difference. The $O(E)$ is quite large.

Now some more interpretations on the properties listed on the floor **G** and **I** are given. The definition of $O(E)$ immediately suggests to consider a finite dimensional analogue. Roughly speaking, choose a finite, say n, dimensional subspace E_n of E. Let g be a linear transformation on E such that its restriction to E_n is an orthogonal transformation, that is in the group $O(n)$ and is identity on the orthogonal complement $E^\perp$. The collection of such transformations is isomorphic to the orthogonal group $O(n)$, let it be denoted by O_n, which is a subgroup of $O(E)$.

To get more concrete analogue of the harmonic analysis arising from finite dimensional Lie group, we prefer to take the rotation group $SO(n)$. Then, we shall introduce a group G_∞, isomorphic to a limit of $SO(n)$, and is a subgroup of $O(E)$, unfortunately we cannot define the determinant in the infinite dimensional case. Take a complete orthonormal system $\{\xi_n\}$ such that $\xi_n \in E$ for every n. The trick is as follows. Take $\xi_j, 1 \le j \le n$, and let E_n be the subspace of E spanned by them. Define G_n by

$$G_n = \{g \in O(E);\ g|_{E_n} \in SO(n),\ g|_{E_n^\perp} = I\}.$$

Obviously, it holds that

$$G_n \cong SO(n).$$

Now the subgroup G_∞ of $O(E)$ can be defined by taking the projective limit of G_n:

$$G_\infty = \text{proj.}\lim_{n\to\infty} G_n.$$

The group G_∞ is certainly infinite dimensional, however, by definition, each member of the group G_∞ can be understood as a transformation that will be approximated by the finite dimensional rotations. The more rigorous and exact interpretation will be given in the next section, where we compare the characteristics of G_∞ with those of the Lévy group introduced there.

One can now understand that the harmonic analysis arising from the infinite dimensional rotation group would be approximated by the finite dimensional analysis. The part **G**, that indicates the ground floor and the first floor **I** contain those which can be approximated by finite dimensional calculus by some method. We claim that the second floor **II** and the third floor **III** exhibit essentially infinite dimensional calculus; namely they cannot be well studied only by using finite dimensional approximations.

Just a short note for the readers: we are going to investigate some details of the relationship between measure and group, i.e. white noise and rotation group, which might be the case only for Gaussian case. This is not quite true, of course. Another typical example for which such a relationship can be seen between Poisson noise and symmetric group. We shall discuss in Chapter 7 and other sections, as well.

At present the following topics are worth to be mentioned.

1) Irreducible unitary representations of $O(E)$. There are many possibilities of introducing unitary representations. It is noted that, unlike finite dimensional Lie group, the dimension of the space on which a unitary representation of $O(E)$ is defined would be infinite. This is one of the reasons why various kinds of representation can be accepted and indeed constructed exactly.

2) Laplacian. The (finite dimensional) spherical Laplacian tends to the infinite dimensional Laplace-Beltrami operator Δ_∞ which is acting on (L^2). There are other Laplacians; in fact, more significant Laplacians are defined and they play different roles in our analysis, respectively. These facts will be discussed in Section 5.8. We also discuss related topics in Chapter 9, in particular Section 9.4.2.

3) The space E spanned by the entries may correspond, in the case $O(E)$, to the space of the span of Fourier-Hermite polynomials, and they are eigenfunctions of the Laplacian Δ_∞ mentioned in 2) above.

4) Although the complex Hilbert space (L^2) is not quite like $L^2(G, m)$, m being the left or right Haar measure of a Lie group $\mathcal{G}$, still one can see some analogy between them. The former is more like an L^2 space over a symmetric space.

5) There is the Fourier-Wiener transform defined on (L^2). It is given by analogy with the ordinary Fourier transform acting on $L^2(R)$, but essential modification is necessary; for one thing the basic measure is not Lebesgue but Gaussian.

These topics will be discussed in the respective sections and should be kept in mind.

Note. The reader is recommended to see Fig. 3 in our monograph[71] page 30.

5.5 The Lévy group, the Windmill subgroup and the sign-changing subgroup of $O(E)$

The group that is going to be defined was first introduced by P. Lévy in the monograph[100] and systematic approach has been done in the book[103] as a tool from functional analysis. Since we recognize the significance of this group also in white noise analysis, we modified the definition slightly, and we wish to find the important roles of this group in white noise analysis.

Let π be an automorphism of $Z_+ = \{1, 2, \cdots\}$. Fix a complete orthonormal system $\{\xi_n\}$ in $L^2(R)$ such that ξ_n is in E for every n.

Let π be an automorphism of Z_+. Then, a transformation g_π of $\xi \in E$ is defined by

$$g_\pi : \xi = \sum_1^\infty a_n \xi_n \to g_\pi \xi = \sum_1^\infty a_n \xi_{\pi(n)},$$

where $g_\pi \xi$ is found in $L^2(R)$.

Define the *density* $d(\pi)$ of the automorphism π by

$$d(\pi) = \limsup_{N\to\infty} \frac{1}{N} \# \{n \le N; \pi(n) > N\}.$$

Set

$$\mathcal{G} = \{g_\pi ; d(\pi) = 0, g_\pi \in O(E)\}. \tag{5.5.1}$$

Obviously, the collection $\mathcal{G}$ forms a subgroup of $O(E)$. It is a discrete and infinite group.

Definition 5.2 The group $\mathcal{G}$ is called the *Lévy group*.

We want to show that the Lévy group has quite a different character from those groups that can be approximated by finite dimensional groups. One of the tools to discriminate $\mathcal{G}$ from familiar group is the *average power* $a.p(g)$ of a member $g = g_\pi$ of $O(E)$. It was introduced for the benefit of those who want to see how an essentially infinite dimensional rotation (transformation) looks like.

We continue to fix the complete orthonormal system $\{\xi_n\}$. Now define the *average power* $a.p(g_\pi)$ by

$$a.p(g)(x) = \lim \sup_{N\to\infty} \frac{1}{N} \sum_1^N \langle x, g\xi_n - \xi_n \rangle^2. \tag{5.5.2}$$

Definition 5.3 If $a.p(g)(x)$ is positive almost surely (μ), then we call g_π *essentially infinite dimensional.*

Quite contrary to this case, if $a.p(g_\pi)(x) = 0$ almost surely, then g_π is tacitly understood to be *approximated* by the finite dimensional rotations.

Remark 5.5 *Sometimes we use the expectation of $a.p(g)$ instead of $a.p(g)$ itself to indicate the same property.*

It is necessary to give some interpretation of the meaning of essentially infinite dimensional rotation mentioned above. By definition, $a.p(g)$ is the Cesàro limit (average) of the square of the distance $|\langle x, g\xi_n\rangle - \langle x, \xi_n\rangle|$. Hence, if $a.p(g)$ is positive, the distance should be larger than some $\epsilon > 0$ for infinitely many n. Intuitively speaking, this means that infinitely many coordinates $\langle x, \xi_n\rangle$ change significantly. Namely, such a change of coordinates cannot be approximated by any change of finite dimensional directions.

We know that for many g in the Lévy group, say $g = g_\pi$,

$$a.p(g_\pi)(x) = \limsup_{N\to\infty} \frac{1}{N}\sum_1^N \langle x, \xi_{\pi(n)} - \xi_n\rangle^2 > 0$$

holds.

We can therefore see that there are many members in the Lévy group (see Example 5.1 below) that are essentially infinite dimensional.

Example 5.1 An example of a member of the Lévy group $\mathcal{G}$.

Let π be a permutation of positive integers defined by

$$\pi(2n-1) = 2n, \quad \pi(2n) = 2n-1, \quad n = 1, 2, \cdots.$$

Fix a complete orthonormal system $\{\xi_n\}$ in $L^2(R)$ such that every ξ_n is in E. For $\xi = \sum a_n\xi_n$ define $g_\pi\xi$ by

$$g_\pi\xi = \sum a_n\xi_{\pi(n)}.$$

This is, as it were, a pairwise permutation of the coordinate vectors. By actual computation, we can easily obtain $a.p(g_\pi) = 2$ almost surely. Hence,

g_π is an essentially infinite dimensional rotation. Obviously, g_π belongs to $O(E)$.

Example 5.2 This is an example of a rotation of E, but it is not part of the Lévy group.

Let N be the set of natural numbers and let I_n be the subset of N consisting of $2^n, \cdots, 2^{n+1} - 1$. Then

$$N = \sum_{n=0}^{\infty} I_n.$$

Take I_n for which we define a permutation π of members in I_n in such a manner that

$$2^n + k \to 2^{n+1} - 1 - k, \qquad k = 0, 1, ..., 2^{n-1} - 1.$$

Then, I_n is arranged in a reflected order. Integers as many as 2^{n-1} are reflected. Hence, we have

$$\limsup_{N \to \infty} \frac{1}{N} \# \{ j \leq N \ \& \ \pi(j) > N \} = \frac{1}{2}.$$

Thus, this π does not satisfy the requirement $d(\pi) = 0$. If the basic nuclear space E is taken to be the Schwartz space S, then the g_π is in $O(S)$, but not in the Lévy group.

It is interesting to note that there should be an intimate connection between the Lévy group and the Lévy Laplacian as we shall see later.

The Windmill subgroup.

There is another subgroup of $O(E)$ that contains essentially infinite dimensional rotations. It is a windmill subgroup $\mathcal{W}$, which is defined in the following manner. In fact, what we are going to define is a generalization of Example 5.1.

Take E to be the Schwartz space S and take an increasing sequence $n(k)$ of positive integers satisfying the condition

$$(n(k+1) - n(k)) \frac{n(k+1)}{n(k)} \leq K, \quad (K > 1).$$

Let $\{\xi_n, n \geq 0\}$ be the complete orthonormal system in $L^2(R)$ such that ξ_n is the eigenfunction of A defined in Section 2.3, Example 2.1, such that

$$A\xi_n = 2(n+1)\xi_n.$$

Let the system be fixed. Denote by E_k the $(n(k+1) - n(k))$-dimensional subspace of $E(= S)$ that is spanned by $\{\xi_{n(k)+1}, \xi_{n(k)+2}, \cdots, \xi_{n(k+1)}\}$.

Let G_k be the rotation group acting on E_k. Then, $\mathcal{W} = \mathcal{W}(\{n(k)\})$ is defined by

$$\mathcal{W} = \bigotimes_k G_k.$$

It is easy to prove

Proposition 5.2 *The system*

$$\mathcal{W} = \mathcal{W}(\{n(k)\})$$

forms a subgroup of $O(S)$ and contains infinitely many members that are essentially infinite dimensional.

Proof is given by evaluating the norm $\|g\xi\|_p, g \in \mathcal{W}$ and by using the requirement on the sequence $n(k)$.

Definition 5.4 The subgroup $\mathcal{W}$ is called a *windmill subgroup.*

Fig. 5.2. Modern windmills in Netherlands

The sign-changing subgroup $\mathcal{H}$.

The group $\mathcal{H}$ is introduced in the following steps:

1) Take $t \in (0, 1]$. Denote the binary expansion of t by

$$t = \sum_{1}^{\infty} \eta_n(t) 2^{-n},$$

where $\eta_n(t) = 0$ or 1.
To guarantee the uniqueness of the determination of $\eta_n(t)$, we define

$$\eta_n(1) = 1 \quad \text{for every } n,$$

and for $p < 2^k$, with $k > 0$,

$$\eta_n(p2^{-k}) = 0$$

for every $n > k$.

2) Set $\epsilon_n(t) = 2\eta_n(t) - 1$. Then, g_t is defined by

$$g_t : \xi = \sum_{1}^{\infty} a_n \xi_n \quad \mapsto \quad \sum_{1}^{\infty} a_n \epsilon_n(t) \xi_n,$$

where $\{\xi_n\}$ is a fixed complete orthonormal system in $L^2(R)$.

3) Every g_t belongs to the group $O(E)$, since it is a continuous linear transformation on E and preserves the L^2-norm, for every t. The collection $\mathcal{H} = \{g_t, t \in (0\ 1]\}$ forms a group, and hence it is a subgroup of $O(E)$, where the product $g_t g_s$ is defined in the usual manner $(g_t g_s)\xi = g_t(g_s \xi)$ and the result is a transformation denoted by $g_{\phi(t,s)}$. Obviously, $\mathcal{H}$ is Abelian: $\phi(t, s) = \phi(s, t)$.

4) Since g_1 is the identity, we have $\phi(t, 1) = \phi(1, t) = t$. By definition, we have

$$\phi(t, t) = 1, \quad \text{i.e.} \quad g_t^2 = e \text{ (identity)}.$$

5) There exists a member $g_t \in \mathcal{H}$ such that the average power of g_t is positive. We may say that the subgroup $\mathcal{H}$ itself is essentially infinite dimensional subgroup of $O(E)$.

Remark 5.6 *It may be interesting to determine the exact form of the function $\phi(t, s)$, but now we do not go into details.*

Definition 5.5 The subgroup $\mathcal{H}$, defined above, is called a *sign-changing subgroup* of $O(E)$.

A significance of the sign-changing subgroup is that it has a connection with certain transformations of sample functions (in fact, generalized functions) of white noise. Indeed, it describes some intrinsic properties of white noise. In other words, the group is interesting also from the viewpoint of the generalized harmonic analysis of stochastic processes. One of the typical examples is given in what follows.

Application of the sign changing group.

Consider Lévy's construction of white noise (in reality the idea is the same as that of Brownian motion, see Lévy[102]) as has been explained in Section 2.1 formula (2.1.7). There i.i.d. (independent identically distributed) standard Gaussian random variables Y_n are involved. Now such a system of i.i.d. random variables is explicitly given on a white noise space (E^*, μ) as follows.

Fix a complete orthonormal system $\{\xi_n\}$. The sign changing group $\mathcal{H}$ consists of members that change the sign of each ξ_n. Recall the action of $g_s \in \mathcal{H}$. The adjoint g_s^* acts on E^* and $g^*\mu = \mu$ holds.

Another notion to be reminded is that on the measure space (E^*, μ) of white noise the system $\{\langle x, \xi_n \rangle\}$ forms an i.i.d. random variables, each of which is subject to the standard Gaussian distribution. We set

$$Y_n' = Y_n'(x) \equiv \langle x, \xi_n \rangle.$$

Then, we have

$$\{Y_n'(x), n \geq 1\} \ \sim \{Y_n'(g_s^* x), n \geq 1\},$$

where $\sim$ means that both systems have the same probability distribution, which is standard Gaussian.

We are now ready to discuss the Lévy construction of a Brownian motion as in Section 2.1. Let us review the method to make clear the situation.

Take the unit interval $[0,1]$ to be the parameter space. Start with $\{X_1'(t)\}$ given by

$$X_1'(t) = tY_1', \ t \in [0,1].$$

The sequence of process $\{X_n'(t), t \in [0,1]\}$ is formed by induction as in Section 2.1.

There is one thing to be noted; namely, $Y_n'(x)$ is defined on $(E^*, \mu) <$ so that the action g_s^* in question is applied to x.

It is easy to see that the sequence $X_n'(t), n \geq 1$, is consistent in the sense that

$$E(X_{n+1}'(t)|\boldsymbol{B}_n(Y')) = X_n'(t), \ t \in [0,1],$$

where $\boldsymbol{B}_n(Y')$ is the sigma-field generated by the $Y_k'(x), \ k \leq 2^{n-1}$. Hence, we can prove that there exists the uniform limit of the $X_n'(t)$. The limit is denoted by $\tilde{X}(t)$. Further, it is proved that the $\tilde{X}(t)$ has independent increments and that

$$E(|\tilde{X}(t) - \tilde{X}(s)|^2) \leq |t-s|,$$

and

$$Cov(\tilde{X}(t), \tilde{X}(s)) = t \wedge s.$$

Letting the white noise variable x revive, it can be shown that

$$\lim_{n\to\infty} X_n'(t,x) = X(t,x),$$

exists and is equal to $\tilde{X}(t)$, a.e. (μ).

Summing up, we have proved the following proposition.

Proposition 5.3 *The processes $X(t,x), t \in [0,1]$ and $X(t, g_s^*), t \in [0,1]$, for any s, are the same which is a Brownian motion.*

Observation

We now make some observation on the sequence $X_n(t)$ that approximates a Brownian motion. It is continuous and piecewise differentiable almost surely. Take the time derivative $X_n'(t)$, which is defined on T_n^c. At each $t \in T_n$ the $X_n'(t)$ has discontinuity of the first kind, that is, there are jumps, the scale of which corresponds to Y_k for the k determined in the definition of $X_n(t)$. Such a sequence of jumps on the set T_n can be thought of as an approximation of a white noise. Thus, we can see intuitive picture of white noise that creates independent infinitesimal Gaussian random variables at each instant t in T_0. Note that the degree of approximation is *uniform* in t over the time interval $[0,1]$. This fact is significant.

Take a transformation $g_a \in \mathcal{H}$, where $a \in [0,1]$. Define $Y_n^a(x) = Y_n(g_a^* x), n \geq 1$. Then, we are given another Brownian motion $X^a(t), t \in [0,1]$. As a result, a transformation of white noise is determined.

With this transformation g_a a U-functional $U(\xi)$ goes to $U(g_a\xi)$ as is easily calculated:

$$
\begin{aligned}
C(\xi)\int \exp[\langle x, \xi\rangle]\varphi(g_a^* x)d\mu(x) &= C(\xi)\int \exp[\langle g_a^* y, \xi\rangle]\varphi(y)d\mu(y), \ (g_a)^* = g_a^{-1} \\
&= C(\xi)\int \exp[\langle y, g_a\xi\rangle]\varphi(y)d\mu(y) \\
&= U(g_a\xi),
\end{aligned}
$$

where $C(\xi) = \exp[-\frac{1}{2}\|\xi\|^2]$, to which $C(g_a\xi)$ is equal.

Proposition 5.4 *The differential operators ∂_t are transformed with the factor g_a under the action of sign changing group.*

Incidentally, it is noted that the Lévy Laplacian, which will be introduced later, is invariant under the sign changing group.

5.6 Classification of rotations in $O(E)$

There is a classification of rotations that are members of the infinite dimensional rotation group $O(E)$. Actually, they are classified into two parts depending on the choice of the basic nuclear space.

(1) **Class I** : Rotation groups based on the coordinate vectors.
(2) **Class II** : Whiskers and related subgroups.

(1) Class I.

There are rotations that are defined with the help of the complete orthonormal system $\{\xi_n\}(\subset E)$ which is fixed in each choice of E. The parameter space can be arbitrary.

Members in $G_\infty(\equiv \text{proj.}\lim G_n)$ are in this class, where $G_n \cong SO(n)$. By definition, we can see that G_∞ can be approximated by the finite dimensional transformations.

The Lévy group, the Windmill subgroup and the sign-changing group are all consisting of rotations in class I. It is noted that they are all essentially infinite dimensional subgroups of $O(E)$.

The definitions of those groups are not always coordinate-free. In particular, the Lévy group depends on the analytic property of the ξ_n. Let the parameter space be taken to be $[0,1]$. It is convenient to assume that the given complete orthonormal system ξ_n in $L^2([0,1])$ is *equally dense.*

We have to pause here to give the definition of *equally dense.* (Ref. P. Lévy[103].) This notion is defined as follows.

Definition 5.6 A complete orthonormal system in $L^2([0,1])$ is said to be equally dense, if the following two conditions are satisfied.

i) $\frac{1}{N}\sum_1^N \xi_n(u)^2 \to 1$, *a.e.* as $N \to \infty$,
ii) $\frac{1}{N}\sum_1^N \xi_n(u)\xi_n(v) \to 0$, a.e. $(u,v), u \neq v$, as $N \to \infty$.

It is noted that the Lévy group has been rooted in the classical functional analysis on $L^2([0,1])$. In particular, it is defined in connection with the Lévy Laplacian. Similar consideration was already made when the windmill subgroup was introduced in the last section. Now we have to assume that a complete orthonormal system is equally dense, when the Lévy group is discussed.

Example 5.3 The following well-known example of complete orthonormal system in $L^2([0,1])$ is equally dense.

$$1,\ \sqrt{2}\sin 2n\pi t,\ \sqrt{2}\cos 2n\pi t,\quad n = 1,2,\cdots.$$

(2) Class II.

A transformation $g \in O(E)$ that comes from a diffeomorphism of the parameter space, like R^d, is classified as class II. For example, the reflection

w with respect to the origin of R^d is a member of $O(D_0(R^d))$ and belongs to class II. The space $D_0(R^d)$ involves C^∞-function over $R^d \bigcup\{\infty\}$, namely it is isomorphic to the nuclear space $C^\infty(S^d)$, where S^d is the d-dimensional sphere.

We are much interested in *whisker* denoted by $g_t, t \in R$. It is a continuous one-parameter subgroup of $O(E)$ that is defined by a one-parameter group of diffeomorphisms $\psi_t(u), t \in R, u \in R^d$, such that

$$g_t\xi(u) = \xi(\psi_t(u))\sqrt{|\psi_t'(u)|}, \quad \psi_t'(u) : \text{Jacobian}, \tag{5.6.1}$$

where

$$\psi_t\psi_s = \psi_{t+s}$$

holds.

We further assume continuity in t of the product of the ψ_t. With a suitable assumptions on ψ_t, it can be proved that the g_t defined above is a whisker. It is a subgroup belonging to class II.

Typical and good examples of a whisker are listed below.

1) The shift. Let $e_j, 1 \le j \le d$, be an orthonormal base of R^d. Define S_t^j by

$$S_t^j\xi(u) = \xi(u - te_j).$$

Then, obviously $S_t^j, 1 \le j \le d, t \in R$, is a whisker. Each S_t^j is called a *shift*.

The shift is extremely important in probability theory and also in other fields. For instance, if $d = 1$, the adjoint $S_t^* = T_t, t \in R$, is a *flow* (after S. Kakutani, it is called a flow of Brownian motion) on the measure space (E^*, μ). Its spectrum is countably Lebesgue on $(L^2) \ominus \{1\}$, in particular, it is ergodic. And so forth.

We wish to find other interesting whiskers. In view of the above fact, we have made the plan of search to use commutation relations of new whiskers with the shift; of course the simpler the better. With this idea we have discovered significant whiskers. See [40] Chapter 5. We list some of them below, although the commutation relations will be shown in the next chapter together with other transformations.

2) Isotropic dilation. The isotropic dilation $\tau_t, t \in R$, is defined by

$$\tau_t \xi(u) = \xi(ue^{\lambda t})e^{td/2}.$$

This is another important whisker.

3) With a special choice of the basic nuclear space to be $O(D_0)$ again, we can define $\kappa_t^j, 1 \leq j \leq d, t \in R$, by using the reflection w, in the following manner

$$\kappa_t^j = wS_t^j w.$$

Put together the above three examples. And we obtain a $(2d+1)$ dimensional Lie group, which is isomorphic to the linear group $SO(d+1,1)$. Its probabilistic role is illustrated in the book[40] Chapter 5. Indeed, the group is a conformal group. We shall be back to this group later in connection with the infinite dimensional unitary group.

Take finitely many whiskers that form a Lie group like in the above example. Then, it is straightforward to apply the well-known technique for finite dimensional Lie groups. As a result, some significant probabilistic properties are investigated with the help of the unitary representation of the Lie group and sometimes by using the algebraic structure (like commutation relations) of the corresponding Lie algebra. Such an approach is more efficient in the case of the unitary group that will be dealt with later in Chapter 6.

We are now ready to discuss the topics on *harmonic analysis* that were announced before. It is the most basic concept in the analysis of functionals for which white noise $\dot{B}(t)$ (or sometimes Poisson noise $\dot{P}(t)$) is taken to be the variable, and the rotation group (symmetric group in Poisson case) plays essential roles.

5.7 Unitary representation of the infinite dimensional rotation group $O(E)$

We shall discuss unitary representations of the infinite dimensional rotation group $O(E)$ in line with harmonic analysis. The definition of the unitary representation and the irreducibility are the same as in the usual case, although our group is infinite dimensional.

First, the Hilbert space (L^2) is taken to define a representation of the group $O(E)$ in question.

For any $\varphi \in (L^2)$ and for $g \in O(E)$ define U_g by

$$(U_g\varphi)(x) = \varphi(g^*x).$$

Then U_g is a unitary operator on (L^2), and the collection $\mathbf{U} = \{U_g, g \in O(E)\}$ forms a group under the product that agrees with the product in $O(E)$. Namely, we have for $g, h \in O(E)$

$$\begin{aligned}(U_g(U_h\varphi))(x) &= (U_h\varphi)(g^*x) = \varphi(h^*(g^*x)) \\ &= \varphi((gh)^*x) = (U_{gh}\varphi)(x).\end{aligned}$$

We can therefore prove that the group $\mathbf{U}$ is isomorphic (algebraically) to $O(E)$, and further the group $\mathbf{U}$ is topologized so as to be isomorphic to the topological group $O(E)$.

For the proof, see the monograph[40].

Theorem 5.3 *i) $\{U_g, g \in O(E); (L^2)\}$ is a unitary representation of the infinite dimensional rotation group $O(E)$.*

ii) The unitary representation above is reduced to $H_n, n \geq 1$, which is the subspace of (L^2) that appears in the Fock space. Then, $\{U_g, g \in O(E); H_n\}$ is an irreducible unitary representation.

More connections to quantum dynamics will be discussed successively in the following sections.

5.8 Laplacians

The next interest is concerned with the Laplacian which will systematically be discussed on the space of white noise functionals.

Discussions on Laplacians related to the white noise theory, including the Lévy Laplacian Δ_L, can be illustrated in terms of quantum dynamics, non-commutative differential geometry and others. We shall, however, without going into these topics, begin with interpretations on Laplacians in a somewhat heuristic order.

Now systematic observations are given.

Laplacian in white noise analysis is not unique. For one thing, the space on which Laplacian acts is infinite dimensional space involving generalized white noise functionals. There have been, so far, at least three different Laplacians. Those Laplacians can be characterized in different manner and their roles are also different, although they satisfy basic requirements as Laplacian like finite dimensional case.

1.a) *The Laplace-Beltrami operator.*

An infinite dimensional version (unfortunately it is not quite a simple generalization, but it is something like a version) of the spherical Laplacian or the Laplace-Beltrami operator is, as is expected, the following operator:

$$\Delta_\infty = \sum_{i=1}^{\infty}\left(\frac{\partial^2}{\partial x_i^2} - x_i \frac{\partial}{\partial x_i}\right). \tag{5.8.1}$$

A direct characterization was made in the paper[37] in somewhat similar spirit to the Peter-Weyl Theorem.

We can easily prove

Proposition 5.5 *The subspace H_n in the Fock space is the eigenspace of the Laplace-Beltrami operator Δ_∞ belonging to the eigenvalue $-n$.*

Proof is given by noting that the Fourier-Hermite polynomial in x_i of degree n is an eigenfunctional of the operator $(\frac{\partial^2}{\partial x_i^2} - x_i \frac{\partial}{\partial x_i})$.

1.b) *Observations from differential geometry.*

We refer to the book by Jost[81], in particular Chapter 2. The Laplace-Beltrami operator Δ (changing the notation) is given by the formula

$$\Delta = dd^* + d^*d,$$

where d is the exterior derivative.

This Laplacian differs from the usual one on R^d by a minus sign, unfortunately. But we have to accept this since it has already been established in geometry.

For a case of Riemannian manifold, the Laplace-Beltrami operator can be expressed by using the Riemannian metric g_{ij} in the following form:

$$\Delta f = \frac{1}{\sqrt{g}} \frac{\partial}{\partial x_i} \left(\sqrt{g} g^{ij} \frac{\partial f}{\partial x_j} \right).$$

If, in particular, a sphere S^d is concerned, an explicit formula of Δf can be obtained. In fact, by using the polar coordinates g_{ij} is given by

$$\begin{aligned} g_{ij} &= 0, \ i \neq j, \\ g_{ii} &= \sin^2 \theta_1 \sin^2 \theta_2 \cdots \sin^2 \theta_{d-1}. \end{aligned}$$

With these g_{ij} we can write a formula for the Δf explicitly.

Unfortunately, the limit of the Laplace-Beltrami operator as $d \longrightarrow \infty$ is not rigorously computed. Intuitively, we use the fact that the uniform measure on the sphere concentrate to the equator as dimension becomes higher, so that the limit is the same as in 1.a).

It is known that the domain of the Laplace-Beltrami operator is the entire space (L^2). The eigensystem is $\{-n, H_n\}, n \geq 0$, that is, the space H_n of homogeneous chaos of degree n is the eigen space belonging to the eigenvalue $-n$. This fact can be proved by the expressions of the Fourier-Hermite polynomials (see the expression (5.5.5)).

In order to discriminate from other Laplacians that will appear in what follows, we write the Laplace-Beltrami operator by Δ_∞.

2) The *Volterra Laplacian* Δ_V (which is, in reality, the same as the *Gross Laplacian* Δ_G).

Recall the white noise derivative ∂_t. The *Volterra Laplacian* denoted by Δ_V is defined by

$$\Delta_V = \int \partial_t^2 dt.$$

The domain of Δ_V is naturally defined in such a way that members of the domain are twice differentiable in ∂_t's and have finite trace. Actual computations are as follows. In the space $\mathcal{F}$ of S-transforms of white noise functionals, the application of Δ_V turns out to be the operation of obtaining the trace of the second order functional derivatives. More explicitly, for S-transform of a homogeneous chaos functional in $\mathcal{F}_n^{(n)} = S(\mathbf{H}_n^{(n)})$ (see

Section 2.5. for the notation $\mathbf{H}_n^{(n)}$) we define $\tilde{\Delta}_V$ by

$$\tilde{\Delta}_V = \int \frac{\delta^2}{\delta\xi(t)\delta\xi(s)}\bigg|_{t=s} dt. \tag{5.8.2}$$

If $\quad (S\varphi)(\xi) = U(\xi) = \int \cdots \int F(u_1, u_2, \cdots, u_n)\xi(u_1)\xi(u_2)\cdots\xi(u_n)du^n,$
then we have

$$\tilde{\Delta}_V U(\xi) = n(n-1)\int \cdots \int F(t, t, u_3, \cdots u_n)\xi(u_3)\cdots\xi(u_n)du^{n-2}dt,$$

where F is assumed that F continuous and the above integral exists. The Volterra Laplacian Δ_V itself is defined by

$$\Delta_V\varphi = S^{-1}\tilde{\Delta}_V S\varphi, \quad \varphi \in \mathbf{H}_n^{(n)}.$$

The domain of Volterra Laplacian is taken to be either $\sum \mathbf{H}_n^{(n)}$ or the Hilbert space $(S)_2$ (for notation see Section 2.6 B).

One can compare our expression of Δ_V with the operator defined by L. Gross[33].

The adjoint operator Δ_V^* to Δ_V is the double creation operator expressed in the form

$$\Delta_V^* = \int (\partial_t^*)^2 dt.$$

It is noted that there is a connection with an infinite dimensional analogue of the *Darboux equation* in the theory of the partial differential equations. Also, the Δ_V is related to the *mean value theorem* for smooth functionals.

Example 5.4 Let $\varphi(x)$ be given by

$$\varphi(x) = \int\int F(u, v) : x(u)x(v) : dudv,$$

where F is continuous and has finite trace. Then, it is in the domain of the

Volterra Laplacian and we have

$$\Delta_V \varphi = \int F(t,t)dt.$$

3) *The Lévy Laplacian* Δ_L.

The Lévy Laplacian was first introduced by Lévy[100] in 1922 as is explained before. Since then, many significant contributions have been made, by himself, Saitô, M.N. Feller, the authors of this book and many others.

To fix the idea, we restrict our attention to the case where the parameter space is taken to be $[0, 1]$. Such a restriction is not essential. The unit interval may be replaced with finite interval without losing basic idea, however in the case infinite time interval, say R, we need minor additional considerations regarding convergence.

Unlike the case of the Laplace-Beltrami operator, where the operator in question is determined by using the Lie algebra of rotation group, our attention is now focused on quadratic forms of differential (annihilation) operations ∂_t's. Before we come to a determination of the Laplacian as a quadratic form of the ∂_t's, let the Lévy Laplacian be given, in our terminology, without any explanation. It is to be of the formula

$$\Delta_L = \int_0^1 \partial_t^2 (dt)^2. \tag{5.8.3}$$

This expression using $(dt)^2$ is rather formal, even strange, but it is illustrated later in the "Observation".

Before we continue, we shall give a quick review of the theory of variational calculus in classical functional analysis.

In the same manner as we defined the partial differential operator, we use the technique in functional analysis with the help of the S-transform.

As a background, we recall a variation of functional and functional derivative up to the second order.

Let $U(\xi), \xi \in E$ be a real-valued functional defined on E. To fix the idea, $U(\xi)$ is assumed to be a U-functional which is the S-transform of a generalized white noise functional. The variation $\delta U(\xi)$ is defined and

Fréchet derivative, assuming its existence, is denoted by $U'(t,\xi)$, where

$$\delta U(\xi) = \int_0^1 U'(t,\xi)\delta\xi(t)dt.$$

The $\delta U(\xi)$ is continuous linear functional of $\delta\xi$, and hence $U'(t,\xi)$ is a generalized function of t for every ξ.

For the second variation, we have more general setup. Let $U(\xi)$ be a U-functional. The second variation, if it exists, has to be of the form (in the sense of Fréchet derivative)

$$\delta^2 U(\xi) = \int\int F(t,s;\xi)\delta\xi(t)\delta\xi(s)dtds, \tag{5.8.4}$$

assuming the existence and its continuity in $\delta\xi^{2\otimes}$. The kernel $F(t,s;\xi)$ is a generalized function of (t,s) for any fixed ξ. Namely, the second order functional derivative is of the form

$$U''_{\xi,\eta}(t,s) = \frac{\delta^2}{\delta\xi(t)\delta\eta(s)}U(\xi) = F(t,s;\xi). \tag{5.8.5}$$

Now assume that $F(t,s;\xi)$ is of a particular form

$$F(t,s;\xi) = f(t)\delta(t-s) + G(t,s,\xi) \tag{5.8.6}$$

where f is continuous and $G(t,s)$ is in $L^2([0,1]^2)$. This corresponds to the classical case; namely the second variation of $U(\xi)$ is expressed in the form

$$\delta^2 U(\xi) = \int_0^1 U''_{\xi\xi}(\xi,t)(\delta\xi(t))^2 dt + \int_0^1\int_0^1 U''_{\xi\eta}(\xi,t,s)\delta\xi(t)\delta\eta(s)dtds.$$

The first term of the right-hand side defines the *singular part* of the second order Fréchet derivative.

Notation The kernel $U''_{\xi\xi}(\xi,t)$ is denoted by

$$U''_{\xi\xi}(\xi,t) = \frac{\delta^2}{\delta\xi(t)^2}U = D_t^2 U. \tag{5.8.7}$$

While, the second term is the *regular part* denoted by $\frac{\delta^2}{\delta\xi(t)\delta\xi(s)}$.

Singularity of F appears only in the form of the first term of (5.8.6). That is $\delta^2 U(\xi)$ is of the normal functional. Applying S^{-1}, the inverse of the S-transform to U, and using the notation

$$\partial_t = S^{-1}(\frac{\delta}{\delta\xi(t)})$$

we have

$$\frac{\delta^2}{\delta x(t)\delta x(s)}\varphi(x) = \int_0^1 \partial_t^2\varphi(dt)^2 + \int_0^1\int_0^1 \partial_t\partial_s\varphi dtds. \tag{5.8.8}$$

We are now in a position to determine the Laplacian which is closely connected with rotations of E^*. The multiplication by $x(t)$ is expressible as the equation

$$x(t)\cdot = \partial_t + \partial_t^*.$$

The infinitesimal generator $r_{s,t}$ of the rotation on the hyperplane $(x(t), x(s))$ is expressed in the form

$$r_{s,t} = x(s)\cdot\partial_t - x(t)\cdot\partial_s.$$

By using the formula of multiplication, we have

$$r_{s,t} = \partial_s^*\partial_t - \partial_t^*\partial_s,$$

as we briefly discussed before.

Let G be a class of $(S)^*$-functionals φ such that $S\varphi$ has second variation satisfying (5.8.8).

Theorem 5.4 *i) The domain G is a vector space, and the second order variational operator D^2, acting on G is expressed in the form*

$$D^2 = D_s^2 + D_r^2 \tag{5.8.9}$$

where

$$D_s^2 = \int_0^1 f(t)\partial_t^2(dt)^2$$

and

$$D_r^2 = \int_0^1\int_0^1 F(t,s)\partial_t\partial_s dtds, \qquad F(t,s) = F(s,t).$$

ii) D_s^2 *with* $f(t)$ = *constant commutes with the rotation* $r_{u,v}$, $(u,v) \in [0,1]^2$, *while* D_r^2 *does not.*

Proof. i) is a rephrasement of the definition of $\mathcal{D}^2$.

ii) is proved by actual computations to see

$$\left[\int_0^1 \partial_t^2 (dt)^2, r_{u,v}\right] = 0,$$

and

$$\left[\int_0^1 \int_0^1 F(t,s)\partial_t \partial_s dt ds, r_{u,v}\right] \neq 0$$

except trivial cases.

Thus, D_s^2 and D_r^2 can be discriminated.

■

Note. Concerning the sum (5.8.9), there are elementary notes on the discrimination between D_s^2 and D_r^2. In fact, convergent property is different. One can see the property in a simple example and find the reason why they are different. Suppose $\{X_n\}$ is a sequence of i.i.d. random variables subject to $N(0,1)$.

A quadratic form

$$Q = \sum_{i,j=1}^{\infty} a_{ij} X_i X_j, \quad a_{ij} \text{ real},$$

is expressed in the form

$$Q = \sum_{i=1}^{\infty} a_{ii}(X_i^2 - 1) + \sum_{i=1}^{\infty} a_{ii} + \sum_{i \neq j} a_{ij} X_i X_j.$$

If mean square convergence should be guaranteed, we have to assume that $\{a_{ii}\}$ is of trace class, while $\{a_{ij}, i \neq j\}$ (off diagonal) is square summable.

Define an operator $\tilde{\Delta}_L$ acting on functional $U(\xi), \xi \in E$, by

$$\tilde{\Delta}_L = \int_0^1 \frac{\delta^2}{\delta \xi(t)^2} dt,$$

by using the singular part of the second order Fréchet derivative.

This corresponds to D_s^2 by taking $f(t) = 1$ under the S-transform. Now we can give

Definition 5.7 The operator

$$\Delta_L = S^{-1}\tilde{\Delta}_L S$$

is called the *Lévy Laplacian.*

By definition, it is easy to see that the Lévy Laplacian annihilates ordinary (L^2)-functionals, since the kernel function of the integral representation is an ordinary function and has no singularity on the diagonal.

Observations

1. The first term on the right-hand side of the second variation may be viewed as a double integral where the integrand involves delta functions on the diagonal. Thus, some interpretation is given to D_s^2, the first term of equation (5.8.9), although it looks like an unusual notation involving $(dt)^2$. Intuitively speaking one dt annihilates singularity with factor $\frac{1}{dt}$ when ∂_t is applied to a generalized functional, and another dt is used for integration.
2. One dt may be considered as the effect to take mean. Roughly speaking, if $[0,1]$ is divided into n subintervals, then Δ_t is $\frac{1}{n}$ so that in the limit we must multiply dt to express the mean.
3. The Laplacians Δ_V and Δ_L commute with the notation $\gamma_{s,t}, (s,t) \in R^2; s \neq t$. This is one of their characteristics.

The domain of the Lévy Laplacian is now understood to be a collection of generalized white noise functionals such that the second variation of their S-transforms involve the terms of the form $U''(\xi, t)$ that is integrable in t. A rigorous definition will be given after the following examples.

Example 5.5 Integrals of the quadratic polynomials : $\dot{B}(t)^2$: in $\dot{B}(t)$'s such as

$$\varphi_1(x) = \int_0^1 f(t) : x(t)^2 : dt \left(= \int_0^1 f(t) : \dot{B}(t)^2 : dt \right),$$

where f is continuous, are involved in the domain of the Lévy Laplacian. It is easy to see

$$\Delta_L \varphi_1(x) = 2 \int_0^1 f(t)dt.$$

Example 5.6 The Gauss kernels with finite time domain like

$$\varphi_2(x) = N_2 \exp\left[c \int_0^1 : x(t)^2 : dt\right],$$

and, more generally,

$$\varphi_3(x) = N_3 \exp\left[c \int_0^1 f(t) : x(t)^2 : dt\right], \quad f \text{ is continuous}$$

are in the domain of Δ_L.

If $\psi(x)$ is in (S), then the product $\psi(x)\varphi_3(x)$ makes sense. The product, let it be denoted by $\varphi_4(x)$, belongs to the domain of the Lévy Laplacian, and we have

$$\Delta_L \varphi_4(x) = 2 \int_0^1 f(t)dt \cdot \varphi_4(x).$$

It is now the time to discuss the domain $\mathcal{D}(\Delta_L)$. It is a subspace of $(S)^*$ or of $(L^2)^-$, members of which have S-transforms with second order functional derivatives, that are of the form (5.8.6). Obviously, we have

Proposition 5.6 *The domain $\mathcal{D}(\Delta_L)$ forms a vector space over the algebra (S).*

Another expression of the Lévy Laplacian comes from the formula

$$\Delta_L' = \lim \frac{1}{N} \sum_1^N \frac{\partial^2}{\partial \xi_n^2},$$

acting on U-functionals, where $\{\xi_n\}$ is a complete orthonormal system in $L^2([0,1])$. In fact, $\frac{\partial}{\partial \xi}$ is the Gâteaux derivative. This is the Cesàro limit of the second order partial derivative in ξ_n-direction, and it is viewed as an average of finite dimensional Laplacians. This means the definition of Δ_L' is reasonable. Indeed, this expression appeared in Lévy's theory of functional analysis[100].

Now one may ask if Δ_L' is in agreement with the $\tilde{\Delta}_L$ acting on U-functionals. This is true under some restriction, if a complete orthonormal system $\{\xi_n\}$ is equally dense. (See Definition 5.6.)

Here we have to make a remark. As is seen below, we may use the Gâteaux derivative to get Laplacian for some favorable cases. It is, however, important that the $\{\xi_n\}$ is chosen to be equally dense. In addition, we understand that Δ_L' is based on countably many coordinates, though the property that equally dense is assumed. While, the original definition base on continuously many coordinates, say $\dot{B}(t), t \in [0,1]$, which are equally weighted. The viewpoint is very much different. In white noise analysis, we often meet such a difference, where profound considerations are necessary. We know that the latter is more significant as is recognized in many places.

For a typical case, say for a homogeneous chaos of degree n, a favorable example of functional $U(\xi)$ is such that

$$U(\xi) = \int_0^1 \cdots \int_0^1 F(u_1, u_2, \cdots, u_k) \\ \xi(u_1)^{n_1}\xi(u_2)^{n_2}\cdots\xi_k(u_k)^{n_k} du_1 du_2 \cdots du_k,$$

where $\sum_1^k n_j = n$, with $n_j \geq 2$, and where $F(u_1, u_2, \cdots, u_k)$ is smooth.

Then, it is easy to prove that for $U = S\varphi$ we have

$$\tilde{\Delta}_L U = S\Delta_L \varphi,$$

which is equal to

$$\sum_j n_j(n_j - 1)\int_0^1 \cdots \int_0^1 F(u_1, \cdots, u_j, \cdots, u_k) \\ \xi(u_1)^{n_1} \cdots \xi(u_j)^{n_j-2} \cdots \xi(u_k)^{n_k} du_1 \cdots du_j \cdots du_k$$

where φ is normal if the kernel function of each φ_n in the chaos expansion of φ has singularities only on the diagonal and is of trace class. (Note that the kernel F above is not yet symmetrized.)

For functionals $\varphi_1(x)$ and $\varphi_2(x)$ in the above examples, we can easily show

$$\Delta_L \varphi_1(x) = 2 \int_0^1 f(t)dt,$$
$$\Delta_L \varphi_2(x) = \frac{2c}{1-2c}\varphi_2(x).$$

For the second equation, we may say that the Gauss kernel is an eigenfunctional of the Lévy Laplacian.

Unfortunately this Laplacian Δ_L is not self-adjoint in the usual sense in the domain that is a subspace of $(S)^*$, however K. Saito and A. Tsoi[134] constructed a suitable space wide enough where Δ_L becomes self-adjoint. Another approach to define the adjoint Δ_L^* is given by Si Si[149] in connection with unitary representation of the infinite symmetric group that is related to the invariance of Poisson noise.

Remark 5.7 *The Lévy Laplacian has close connection with the Lévy group. Direct characterization of Δ_L by the Lévy group $\mathcal{G}$ is still not quite satisfactory, however, Theorem 5.4 tells us that $r_{s,t}$'s, the infinitesimal generators coming from $\mathcal{G} \vee G$ (see Diagram 1-1,1-2), characterizes Δ_L in a sense.*

Remark 5.8 *Laplacians Δ_V and Δ_L share the roles in the second variation of a white noise functional. The former comes from the regular part, while the latter is determined by the singular part.*

Proposition 5.7 *Let E' be the image of the domain of Δ_L by the S-transform. Then $\tilde{\Delta}_L$ is a derivation on E'.*

Proof. Let $U_i(\xi), i = 1, 2$, be in E, and assume that their product $U_1(\xi)U_2(\xi)$ is also in E. Then the Leibniz rule is applied to the second variation of the product. Taking the singular parts, we have

$$\tilde{\Delta}_L(U_1(\xi)U_2(\xi)) = (\tilde{\Delta}_L U_1(\xi))U_2(\xi) + U_1(\xi)(\Delta_L U_2(\xi)).$$

Lévy Laplacian on $(S)^*$ with parameter set R instead of $[0, 1]$ should also be discussed. It is, however, not easy to have good connection with classical functional analysis, since it seems difficult to have an extended notion of equally dense complete orthonormal system. Still we can define Δ_L by using the same expression as,

$$\Delta_L = \int_{-\infty}^{\infty} (\partial_t)^2 (dt)^2,$$

although the domain of integration is replaced by R. The same interpretation for the case of some general parameter set is given to have a similar formula.

Finally, we note an important remark.

Remark 5.9 *To define Laplacians Δ_V and Δ_L we have used ∂_t's, Fréchet derivative. This means that each t defines the differential operator ∂_t uniformly in t. In fact, we have partial differential operators for all directions as many as continuum (formally speaking). This is quite different from the method of using Gâteaux derivatives for the directions as many as countable, when t disappears.*

Chapter 6

Complex white noise and infinite dimensional unitary group

6.1 Why complex?

Here are keywords that explain the reasons why complexification in analysis is recommended.

1) complexified versions of real objects are often useful as in the classical complex analysis, that is complex function theory; the same for the white noise analysis.
2) character of abelian group and Pontryagin duality: duality is a fundamental concept in mathematics. Unitary representation of a complexified rotation group, that is unitary group, can appear in a natural manner.
3) the Fourier transform of complex-valued function(al)s can be generalized to infinite dimensional case. It plays more than the ordinary Fourier transform does in the finite dimensional analysis.
4) analysis on the half space, the modular structure can be discussed, and
5) holomorphic functionals enjoy various useful analytic properties, Laplacian and others; the same in our infinite dimensional case.

It is possible to have the infinite dimensional rotation group $O(E)$ complexified to obtain the infinite dimensional unitary group. We also have complexification of Lie algebras associated with the group $O(E)$. Thus, the complex version naturally comes into our view.

It is known that, in the case of a finite dimensional real Lie group, the complexified group is easier to be investigated rather than a given real Lie group itself. This fact is still true, although not for everything, in the infinite dimensional case. We shall, therefore, be playing a similar game in the infinite dimensional rotation group in question.

To this end, we first complexify a white noise to have a complex white noise. Then, we come to an infinite dimensional unitary group acting on a complexified nuclear space. It is convenient to take the real Schwartz space $\mathcal{S}$ as the basic nuclear space E. Once it is complexified to have $\mathcal{S}_c$, then the Fourier transform becomes one of the important members of the infinite dimensional unitary group, and it plays a key role in our study.

6.2 Some background

A complex Gaussian random variable Z is a complex-valued variable such that $Z = X + iY$, where X and Y are real and independent, and they are subject to the same Gaussian distribution with mean 0. This choice of expression of complex Gaussian random variable is fitting for finding analogous properties between real and complex Gaussian systems.

We are now ready to introduce necessary concepts rigorously and discuss the main properties.

Let E_c and E_c^* be the complexifications of a real nuclear space E and its dual space E^*, respectively: More precisely, we define complexified spaces:

$$\begin{aligned} E_c &= E + iE, \\ E_c^* &= E^* + iE^*. \end{aligned}$$

An element ζ of E_c and an element z of E_c^* are written in the form

$$\begin{aligned} \zeta &= \xi + i\eta, \quad \xi, \eta \in E, \\ z &= x + iy, \quad x, y \in E^*, \end{aligned}$$

respectively. The canonical bilinear form $\langle x, \xi \rangle, x \in E^*, \xi \in E$, extends to a continuous sesquilinear form $\langle z, \zeta \rangle, z \in E_c^*, \zeta \in E_c$, that connects E_c and E_c^*. Namely, we have

$$\langle z, \zeta \rangle = (\langle x, \xi \rangle + \langle y, \eta \rangle) + i(-\langle x, \eta \rangle + \langle y, \xi \rangle). \tag{6.2.1}$$

Take white noise measures μ_1 and μ_2 on $E^*(\cong iE^*)$ with variance $\frac{1}{2}$. Let $\mathbf{B}$ be the sigma-field generated by the cylinder subsets of E_c^*, and form a measurable space $(E_c^*, \mathbf{B})$ on which the product measure

$$\nu = \mu_1 \times \mu_2$$

is introduced.

Definition 6.1 The measure space $(E_c^*, \mathbf{B}, \nu)$ is called a *complex white noise.*

Customary z in E_c^* (with measure ν) is also called a complex white noise.

The characteristic functional of $\mu_1 \times \mu_2$ is

$$\begin{aligned} C(\xi, \eta) &= \int\int_{E^* \times E^*} e^{i\langle x, \xi\rangle + i\langle y, \eta\rangle} d\mu_1(x) d\mu_2(y) \\ &= \int_{E^*} e^{i\langle x, \xi\rangle} \mu_1(x) \int_{E^*} ie^{i\langle y, \eta\rangle} d\mu_2(y) \\ &= e^{-\frac{1}{4}(\|\xi\|^2 + \|\eta\|^2)}. \end{aligned} \tag{6.2.2}$$

A *complex Brownian motion* $Z(t), t \geq 0$, is defined by a stochastic bilinear form

$$Z(t)(= Z(t, z)) = \langle z, \chi_{[0,t]}\rangle.$$

Intuitively speaking, the time derivative of $Z(t)$, that is $\dot{Z}(t)$ a sample function of which is identified $z(t)$ and it is also called a complex white noise.

The complex Hilbert space $(L_c^2) = L^2(E_c^*, \mathbf{B}, \nu)$ is the space of functionals of complex white noise. It consists of all $\mathbf{B}$-measurable functionals of z with finite variance, i.e. square integrable functions with respect to the measure ν.

Triviality. We have

$$(L_c^2) \cong (L^2)_x \bigotimes (L^2)_y.$$

The *Fock space* for complex white noise is of the form

$$(L_c^2) = \bigoplus_{n=0}^{\infty} (\bigoplus_{k=0}^{n} H_{(n-k,k)}) \equiv \bigoplus_{n=0}^{\infty} H_n, \tag{6.2.3}$$

where $H_{(n-k,k)}$ is the space spanned by complex Fourier-Hermite polynomials of degree $(n-k)$ in $\langle z, \zeta \rangle$'s and of degree k in $\langle \bar{z}, \bar{\zeta} \rangle$'s. The complex Fourier-Hermite polynomials in complex white noise are defined by using the complex Hermite polynomials, the exact formulas of which are listed in Appendix 3.

As we shall see later, the direct sum decomposition of H_n into the subspaces $H_{(n-k,k)}, 0 \le f \le n$, can be done by the action of the subspace of the infinite dimensional unitary group introduced later.

We are interestd in the subspace H^h of (L_c^2) defined by the algebraic sum

$$H^h = \bigoplus_{n=0}^{\infty} H_{(n,0)}.$$

Typical example $\varphi(z)$ in H^h is defined by a polynomial $p(t_1, \cdots, t_n)$ in t_j's and variables $\langle x, \xi_1 \rangle, \cdots, \langle x, \xi_n \rangle$ in such a way that

$$\varphi(z) = p(\langle x, \xi_1 \rangle, \cdots, \langle x, \xi_n \rangle).$$

Such a polynomial satisfies (formally, but justified easily)

$$\frac{\partial \varphi}{\partial \bar{z}} = 0,$$

namely, it is a *holomorphic polynomial.* We claim that H^h is spanned by holomorphic polynomials.

Many analogous results to those in the holomorphic function theory of one-dimensional variable $z \in C$ can be obtained. They are interesting, however we do not go into details. Readers, who are interested in this direction, are recommended to see the literature[40] Chapter 6.

The space $(L_c^2)^-$ and $(S_c)^*$ of *generalized complex white noise functionals* are defined in the same manner as in the case of Section 2.6; the former is defined by using the Fock space for (L_c^2) and the latter uses the second quantization method.

We are now ready to define the infinite dimensional unitary group. Denote by $U(E_c)$ the collection of all linear transformations g on E_c^* that satisfy the conditions:

1. g is a linear homeomorphism of E_c,
2. g preserves the complex $L^2(R)$-norm:

$$\|g\xi\| = \|\xi\|, \quad \zeta \in E_c.$$

A product $g_1 g_2$ for $g_1,\ g_2 \in U(E_c)$ is defined by

$$(g_1 g_2)\zeta = g_1(g_2\zeta).$$

Under this product $U(E_c)$ forms a group. We introduce the compact-open topology to $U(E_c)$ so as to be a *topological group*, as in the case of $O(E)$.

Definition 6.2 The topological group $U(E_c)$ is called the *infinite dimensional unitary group.*

The adjoint g^* of g in $U(E_c)$ is defined in the usual manner:

$$\langle z, g\zeta\rangle = \langle g^* z, \zeta\rangle,$$

where $\langle\cdot,\cdot\rangle$ is the sesquilinear form defined by (6.2.1).

The collection of such g^*'s forms a group which is denoted by $U^*(E_c^*)$.

Example 6.1 Let g be a multiplication by $e^{i\theta}, \theta \in [0, 2\pi]$.

$$g\xi(u) = e^{i\theta\xi(u)}.$$

Then

$$g^* z = e^{-i\theta} z,$$

which defines a rotation of (x, y)-plane $E^* \times E^*$:

$$(x\ y) \mapsto (x\cos\theta - y\sin\theta, x\sin\theta + y\cos\theta).$$

The characteristic functional is

$$\int\int_{E^*\times E^*} e^{i\langle x\cos\theta - y\sin\theta, \xi\rangle + i\langle x\sin\theta + y\cos\theta, \eta\rangle} d\mu_1(x) d\mu_2(y)$$
$$= \int_{E^*} e^{i\langle x, \xi\cos\theta - \eta\sin\theta\rangle} d\mu_1(x) \int_{E^*} e^{i\langle y, -\xi\sin\theta + \eta\cos\theta\rangle} d\mu_2(y)$$

$$= e^{-\frac{1}{4}\|\xi\cos\theta+\eta\sin\theta\|^2} e^{-\frac{1}{4}\|-\xi\sin\theta+\eta\cos\theta\|^2}$$

$$= e^{-\frac{1}{4}(\|\xi\|^2+\|\eta\|^2)}$$

$$= C(\xi, \eta).$$

Proposition 6.1 *For every $g^* \in U^*(E_c^*)$ we have*

$$g^*\nu = \nu. \tag{6.2.4}$$

Proof. i) As in the case of the rotations, the collection of cylinder subsets of E_c^* is kept invariant under $g \in U(E_c)$, hence follows the **B**-measurability of g^*.

ii) g^*z is represented by $(\frac{1}{2}(g^*z+\overline{g^*z}), \frac{1}{2i}(g^*z-\overline{g^*z}))$. Hence the characteris functional is

$$\int\int e^{i\frac{1}{2}\langle g^*z+\overline{g^*z}\xi\rangle+i\frac{1}{2i}\langle g^*z-\overline{g^*z},\eta\rangle} d\nu$$

$$= \int\int e^{\frac{i}{2}(\langle z,g\xi\rangle+\overline{\langle z,g\xi\rangle})+\frac{1}{2}(\langle z,g\eta\rangle-\overline{\langle z,g\eta\rangle})} d\nu \quad \text{(since } \xi \text{and } \eta \text{ arereal)}$$

$$= \int\int e^{i\langle x,g^*\xi+g^*\eta\rangle+i\langle y,g\bar{\xi}-\bar{g}\eta\rangle} d\mu_1(x)d\mu_2(y)$$

$$= \frac{1}{8}(\|g\bar{\xi}+\overline{g\xi}\|^2+\|g\bar{\eta}-\overline{g\eta}\|^2)$$

$$= e^{-\frac{1}{4}(\|\xi\|^2+\frac{1}{4}\|\eta\|^2)}.$$

This proves the assertion. ∎

We are, therefore, given a untary operator U_g defined by

$$U_g\varphi(z) = \varphi(g^*z), \ \varphi \in (L_c^2).$$

Under the usual product the collection of U_g's forms a topological group that is isomorphic to the group $U(E_c)$. Namely, we are given a unitary representation $\{U_g, g \in U(E_c); (L_c^2)\}$ of the infinite dimensional unitary group $U(E_c)$.

It is noted that the infinite dimensional rotation group $O(E)$ may be identified with a subgroup of $U(E_c)$. In fact, there is a subgroup involving the members such that the restrictions to E belong to $O(E)$. More precisely, let g in $U(E_c)$ be restricted to E, and let the restriction be denoted by $\tilde{g}$. If $\tilde{g}$ is in $O(E)$, and if for $\zeta = \xi + i\eta$ the g acts in such a way that

$$g\zeta = \tilde{g}\xi + i\tilde{g}\eta\,.$$

Obviously, the collection of such g's forms a subgroup of $U(E_c)$ and is isomorphic to $O(E)$.

We also note that, in what have been discussed so far, the parameter space can be taken to be the multi-dimensional Euclidean space or a symmetric space, like S^c, obtained from a Lie group and in addition, we note that there is much freedom to choose a basic nuclear space E_c.

6.3 Subgroups of $U(E_c)$

1) Finite dimensional unitary group $U(n), n \geq 1$ and analogue of the windmill subgroup.

As in the case of taking the subgroup G_n of $O(E)$, we choose an orthonormal system $\zeta_k, k \geq 1$, in E_c. The first n members determine an n-dimensional subspace E_n and the unitary group $U(n)$ acting on E_n is defined. The group $U(n)$ can be embedded in the group $U(E_c)$ as its subgroup. Further, the projective limit $U(\infty)$ of $U(n)$ can also be introduced.

Another generalization of the technique is as follows. Take members in E_c as many as n_k such that they form an orthonormal system. They form a subspace E_{n_k}. We take $E_{n_k}, k = 1, 2, \cdots, m$, which are mutually orthogonal. Thus a subspace $\bigoplus E_{n_k} \subset E_c$ is given. Hence, we can form a product $U(\mathbf{n}) = \bigotimes U(n_k)$ of the unitary groups $U(n_k)$ defined as above, where $\mathbf{n} = (n_1, n_2, \cdots, n_m)$. By letting $m \to \infty$ we are given an analogue of the windmill subgroup $\mathcal{W}$ (see Chapter 5 Section 5.5) of $O(E)$.

2) Conformal group.

If, in the R^d-parameter case, the basic nuclear space E is taken to be $D_{0,c}$, which is isomorphic to complex $C^\infty(S^d)$ (the isomorphism is defined as a generalization of the case $d = 1$ that is discussed in Appendix III, then we are given the conformal group $C(d)$ which is a subgroup of $O(E)$ as was briefly mentioned before Hence, the complex form of $C(d)$, acting on the space $D_{0,c}$, is essentially the same group as in the real case. So, we simply use the same symbol $C(d)$ instead of $C_c(D)$, and call it a *complex conformal group*. It is locally isomorphic to the (real) linear group $SO(d+1,1)$ and is generated by one-parameter subgroups of whiskers as many as $\frac{(d+1)(d+2)}{2}$.

Let us remind the real case (in Chapter 5). The group is generated by four classes of whiskers. The shift S_t^j is the natural extension of the shift, a subgroup of $O(E)$, to the group of operators acting on complex nuclear space. We use the same notation. If $d = 1$, the one-parameter group $S_t^*, t \in R$, defines the *flow of the complex Brownian motion* $\{T_t\}$ with $S_t^* = T_t$.

Other whiskers are the isotropic dilation τ_t, rotations of R^d-space and special conformal transformations. We do not have similar discussions to those in Chapter 5, but we are now interested in the infinitesimal generators of those whiskers.

The generators of the shift, the isotropic dilation , the group of rotations of R^d and the special conformal transformations are, repectively, as follows:

$$\begin{aligned} s_j &= -\frac{\partial}{\partial u_j}, \quad j = 1, 2, ..., d, \\ \tau &= \lambda(u, \nabla) + \frac{d}{2}, \quad r = |u|, \\ r_{j,k} &= u_j \frac{\partial}{\partial u_k} - u_k \frac{\partial}{\partial u_j}, \quad 1 \le j \ne k \le d, \\ \kappa_j &= u_j^2 \frac{\partial}{\partial u_j} + u_j, \quad j = 1, 2, ..., d. \end{aligned}$$

Needless to say, the basic nuclear space E has to be the space D_0, as is explained in Section 5.6. (2), since the reflection w is involved.

Remark 6.1 *Note that in Section 5.6 (2), 2), isotropic dilation is defined, where dilation is restricted only to be isotropic, instead of general dilations acting on R^d. In order to have a finite dimensional Lie algebra (group), we should exclude general dilations.*

3) The Heisenberg group.

From now on, one can see the effective use of complex white noise with emphasis on the role played by the Fourier transform, actually in many places. The basic nuclear space E_c is now specified to be the complex

Schwartz space

$$\mathcal{S}_c = \mathcal{S} + i\mathcal{S},$$

so that the Fourier transform $\mathcal{F}$ itself is a member of $U(S_c)$.

3.1) The gauge transformation $I_t, t \in R$, is defined by

$$I_t : \zeta(u) \longrightarrow I_t\zeta(u) = e^{it}\zeta(u).$$

Obviously I_t is a member of $U(E_c)$, and $\{I_t\}$ forms a continuous one-parameter subgroup. It is periodic with period 2π.

$$\begin{aligned} I_t I_s &= I_{t+s}, \quad t, s \in R, \\ I_{t+2\pi} &= I_t, \\ I_t &\to I \text{ as } t \to 0. \end{aligned}$$

The group $\{I_t, t \in R\}$ is called the *gauge group*, the infinitesimal generator of which is iI, where I is the identity operator. Actually, we have an abelian gauge group that is isomorphic to $U(1) \cong S^1$. We borrow the term gauge group from physics; in fact, it is the simplest gauge group in quantum mechanics. We shall discuss a somewhat general case later in this subsection.

Let the unitary operator U_t be denoted by U_{I_t}. The collection $\{U_t, t \in R\}$ forms a one–parameter unitary group acting on (L_c^2). This group has only point spectrum. The eigenspace belonging to the eigenvalue $-n + 2k$ is $H_{(n-k,k)}$. Hence, the following proposition is proved:

Proposition 6.2 *The space $H_n, n > 1$, is classified, according to the action of I_t, into its subspaces $H_{(n-k,k)}$ for which eigenvalue $-n + 2k$ is associated.*

3.2) The shift S_t^j $1 \le j \le d$.

The shift appears to play the central role in this case, too. The generators have been given in 2).

3.3) Multiplication $\pi_t^j, j = 1, 2, \cdots, d$. Let them be defined to be the conjugate to the shifts via the Fourier transform $\mathcal{F}$:

$$\pi_t^j = \mathcal{F} S_t^j \mathcal{F}^{-1}.$$

Actual expressions are

$$(\pi_t^j \zeta)(u) = e^{itu_j}\zeta(u), \ \ u \in R^d.$$

The infinitesimal generators of the multiplications are denoted by π^j and it is expressed in the form

$$\pi^j = iu_j \cdot .$$

The multiplication acts as a shift on the spectral measure of the shift.

Definition 6.3 The subgroup of the $U(\mathcal{S}_c)$ generated by the gauge group, the shifts and the multiplication is called the *Heisenberg group.*

It should be noted that we have the commutation relation

$$\pi_t S_s = I_{st} S_s \pi_t.$$

In terms of the generators

$$[\pi, s] = iI,$$

which is most significant relation (actually, it is nothing but the *uncertainty principle*).

Gauge transformations (continued)

We can extend the Heisenberg group which shows an invariance of complex white noise. In fact, the group of gauge transformations $\{I_t\}$ extends to a bigger group which has analytic interest.

3.4) Define I_α by

$$I_\alpha : \ \zeta(u) \ \rightarrow \ (I_\alpha \zeta)(u) = e^{i\alpha(u)}\zeta(u),$$

where α is a member of the (real) Schwartz space S. The operator I_α is a unitary operator on $L^2(R)$ and is a member of $U(S_c)$. It is called *S-gauge transformation.* The collection of the S-gauge transformations $\{I_\alpha; \alpha \in S\}$ forms an abelian group under the usual product, and it is a subgroup of $U(S_c)$. In particular, $\{I_{\alpha t}; y \in R\}$ for a fixed α is a one-parameter subgroup of $U(S_c)$ that is not a whisker.

The infinitesimal generator of $I_{\alpha t}$ is $i\alpha$ (which means multiplication by $i\alpha(u)$), where $\alpha \in S$.

We now list the infinitesimal generators of the whiskers listed in 3.1) – 3.4) above.

$$iI; \quad s = -\frac{d}{du}, \quad \pi = iu\,\cdot, \quad i\alpha\,\cdot,$$

where $\alpha \in S$.

Non-trivial commutation relations are

$$[\pi, \alpha] = 0, \ [s, \alpha] = \alpha' \ (\in S).$$

The adjoint operator I_α^* acts on $z \in S_c$ in such a way that $e^{-i\alpha(u)}z(u)$, and

$$I_\alpha^* = e^{-i\alpha(u)}\,\cdot\,, \quad \alpha \in S,$$

is an S^*-*gauge transformation*.

The collection

$$\left\{I_\alpha^* = e^{-i\alpha(u)}\,\cdot\,, \quad \alpha \in S\right\}$$

is called S^*-*gauge transformation group*.

Hereafter we omit $\cdot$ for simplicity.

For the generators discussed so far, we have a theorem.

Theorem 6.1 *i) The generators of the conformal group and the Heisenberg group listed above generate a finite dimensional Lie algebra under the Lie product* $[\ ,\ ]$.

ii) The S^**-gauge transformation acts on* $z = (x, y)$*-space, and* ν *measure is invariant under the action of the adjoint* I_α^*.

The proof of part i) is given by simple computations. As for ii), the characteristic functional guarantees the result.

We stated in i) of the theorem a vague assertion on the Lie algebra. In the case $d = 1$, clear and interesting commutation relations can be seen as we shall see later in this section.

Note. What was discussed in 3.4) above may be said to be a generalization of 3.1), but not quite. Constants are not contained in S.

4) The Fourier-Mehler transforms $\mathcal{F}_\theta$.

Since particularly important roles of the Fourier transform can be seen in the study of complex white noise, we shall further proceed to the fractional powers of the ordinary Fourier transform. Actually, we define a one-parameter group of unitary operators $\mathcal{F}_\theta, \theta \in [0, 2\pi]$ such that $\mathcal{F}_\theta$ is viewed as the $\frac{2\theta}{\pi}$-th (fractional) power of the ordinary Fourier transform, where θ is measured *mod* 2π.

The operator in question is defined by the integral kernel $K_\theta(u,v)$:

$$K_\theta(u,v) = (\pi(1 - \exp[2i\theta]))^{-1/2} \exp\left[-\frac{i(u^2+v^2)}{2\tan\theta} + \frac{iuv}{\sin\theta}\right]. \qquad (6.3.1)$$

It defines an operator $\mathcal{F}_\theta$ by the formula

$$(\mathcal{F}_\theta \zeta)(u) = \int_{-\infty}^{\infty} K_\theta(u,v)\zeta(v)dv, \qquad (6.3.2)$$

where $\theta \neq \frac{1}{2}k\pi,\ k \in Z$.

Definition 6.4 The transformation defined by $\mathcal{F}_\theta$ is called the *Fourier-Mehler transform.*

We now have some observations about the actions of $\mathcal{F}_\theta$. Take the Hermite function:

$$\xi_n(u) = (2^n n! \sqrt{\pi})^{-1/2} H_n(u) \exp[-\frac{u^2}{2}].$$

Then, it is proved that

$$\mathcal{F}_\theta \xi_n(u) = e^{in\theta}\xi_n(u), \quad n \geq 0.$$

With this relationship we can prove that $\mathcal{F}_\theta$ is well defined for every θ (by interpolation), and further

$$\mathcal{F}_\theta \mathcal{F}_{\theta'} = \mathcal{F}_{\theta+\theta'} = \mathcal{F}_{\theta''}, \quad \theta + \theta' = \theta'' \ (mod\ 2\pi).$$

$$\mathcal{F}_\theta \to I, \text{ as } \theta \to 0.$$

Particular choices of θ give

$$\mathcal{F}_{\pi/2} = \mathcal{F}, \quad \mathcal{F}_{(3/2)\pi} = \mathcal{F}^{-1}.$$

Thus, we have obtained a periodic one-parameter unitary group including the Fourier transform and its inverse.

The kernel function K_θ illustrates this fact. Moreover, one can see that the $\mathcal{F}_\theta$ defines a fractional power of the Fourier transform as is shown in what follows.

First we note an identity

$$vK_\theta(u,v) = (u\cos\theta - i\frac{\partial}{\partial u}\sin\theta)K_\theta(u,v).$$

Namely, the multiplication operator "u" is transformed by $\mathcal{F}_\theta$ by the following formula

$$u \;\mapsto\; u\cos\theta + \frac{1}{i}\frac{\partial}{\partial u}\sin\theta.$$

This comes from the above identity.

Similar computation proves that we are given, under $\mathcal{F}_\theta$,

$$\frac{1}{i}\frac{\partial}{\partial u} \;\mapsto\; -u\sin\theta + \frac{1}{i}\frac{\partial}{\partial u}\cos\theta.$$

Symbolically writing, we have established

$$\mathcal{F}_\theta \;:\; \begin{pmatrix} u \\ \frac{1}{i}\frac{\partial}{\partial u} \end{pmatrix} \mapsto \begin{pmatrix} \cos\theta & \sin\theta \\ -\sin\theta & \cos\theta \end{pmatrix} \begin{pmatrix} u \\ \frac{1}{i}\frac{\partial}{\partial u} \end{pmatrix}$$

We are happy to see that there appeared the group $SO(2)$ again.

The infinitesimal generator of $\mathcal{F}_\theta$ is denoted by if and is expressed in the form

$$if = -\frac{1}{2}i(\frac{d^2}{du^2} - u^2 + I).$$

Observing the commutation relations of the generators, so as to have a finite dimensional Lie algebra, either real form or complex form, we are naturally given a new generator σ' which is expressed in the form

$$\sigma' = \frac{1}{2}(\frac{d^2}{du^2} + u^2).$$

We are particularly interested in the probabilistic roles of this operator (generator) in quantum dynamics.

An easy and formal interpretation of σ' is that

$$\frac{1}{i}\partial_t \Psi = \sigma' \Psi$$

is the Schrödinger equation for the *repulsive oscillator.*

For our purpose, it is convenient to take $\sigma = \sigma' + \frac{i}{2}I$, namely we introduce

$$\sigma = \frac{1}{2}(\frac{d^2}{du^2} + u^2 + iI).$$

A one-parameter group with the generator σ can be defined locally in space-time only. It is, however, interesting to discuss the operator σ in connection with the dynamics having a potential of repulsive force.

Lie algebras of infinitesimal generators.

We restrict our attention to the case where $d = 1$ to avoid nonessential complexity in order to observe the significant structure of the Lie algebra generated by the generators that we have obtained so far.

Taking $d = 1$, we list all the generators. For the conformal group there are

$$s = -\frac{d}{du},$$
$$\tau = u\frac{d}{du} + \frac{1}{2},$$
$$\kappa = u^2\frac{d}{du} + u.$$

From now on we come to the world of complex numbers and operators, where one can find not only complexification of the real Lie algebras that have already been given, but also new members having close connection with the Fourier transform.

The Heisenberg group defines the Lie algebra generated by

$$\begin{aligned} &iI, \\ s &= -\frac{d}{du}, \\ i\pi &= iu. \end{aligned}$$

Hereafter, the one-parameter subgroups $I_{\alpha t}$ will be treated separately. One reason lies in the fact that we wish to let the algebra remain within finite dimensional Lie algebras (Lie subgroups).

There are two interesting generators related to the Fourier transform. They are listed again:

$$\begin{aligned} if &= -\frac{i}{2}(\frac{d^2}{du^2} - u^2 + I), \\ \sigma &= \frac{1}{2}(\frac{d^2}{du^2} + u^2 + iI). \end{aligned}$$

We are now ready to consider the algebraic structure of the Lie algebras generated by operators that have appeared so far except for the S-gauge transformations. The structure would reflect probabilistic properties of complex white noise.

Recall the Lie algebra $\mathbf{c}(d)$ of the conformal group. In particular, $\mathbf{c}(1)$ is

$$\mathbf{c}(1) = \{s, \tau, \kappa\},$$

where we use $\{\ \}$ to denote the Lie algebra generated by the members in the curly brackets for convenience. In the case of multi-dimension, say d-dimension, the rotations $\gamma_{j,k}$, $1 \le j, k \le d$, are involved in $c(d)$, but now we skip it.

The Lie algebra of the d-dimensional Heisenberg group is denoted by $\mathbf{h}(d)$. For the present case $d = 1$ we have

$$\mathbf{h}(1) = \{iI, s, \pi = iu\}.$$

Keep the following two concepts in mind.

1. The ordinary Fourier transform $\mathcal{F}$ plays a key role.

2. $\mathcal{F}$ is a member of the unitary group $U(S_c)$.

Then we have naturally been led to the Fourier-Mehler transform $\mathcal{F}_\theta$, the fractional power of the Fourier transform that forms a one-parameter subgroup of $U(S_c)$ with the generator if. Note that we have introduced the operators, first σ', then σ: in fact, by hand for a moment, and later meaningful interpretation is given so that the commutation relations appear in good shape. We have therefore had

$$\sigma = \frac{1}{2}(\frac{d^2}{du^2} + u^2 + iI).$$

Summing up

Proposition 6.3 *Based on the set of operators listed below*

$$iI, s, \pi, \tau, f, \sigma,$$

we have 6-dimensional complex Lie algebra $\mathbf{g}$.

This algebra has the real form as is easily seen.

Table of commutation relations.

For $\mathbf{c}(1)$,

$$[\tau, s] = -s,$$

$$[\tau, \kappa] = \kappa,$$

$$[s, \kappa] = -2\tau.$$

For $\mathbf{h}(1)$,

$$[\pi, s] = I.$$

For the algebra $\mathbf{g}$,

$$[f, s] = \pi, \quad [\sigma, s] = \pi,$$

$$[f, \pi] = s, \quad [\sigma, \pi] = -s,$$

$$[f, \tau] = -2\sigma + iI, \quad [\sigma, \tau] = 2f - I.$$

The algebra $\mathbf{h}(1)$ is an ideal of $\mathbf{g}$ and is the maximum solvable Lie subalgebra. In short we may state

Proposition 6.4 *The ideal* $\mathbf{h}(1)$ *is the radical of* $\mathbf{g}$.

Proof is given by the actual and rather easy computations.

It seems necessary to give some interpretation, from various viewpoints, to the generator κ, which is a member of the algebra associated with the conformal transformations.

[1] The reason why κ has been taken.

i) Obviously κ is a good candidate to be introduced to a class of possible generators expressed in the form $a(u)\frac{d}{du} + \frac{1}{2}a'(u)$ associated with a whisker. We see that when the basic nuclear space E is taken to be D_0, κ is well acceptable. Having had the κ in our class, we have formed the algebra generated by those admissible generators and we see that the algebra is finite dimensional and is isomorphic to $sl(2, R)$. This is really a beautiful result.
ii) Similar to s, κ is transversal to τ. This is a significant property that is typical case among relationships between two members of a Lie algebra. In fact, it defines a flow of the Ornstein-Uhlenbeck process to which the shift is transversal.
iii) The κ is related to the reflection with respect to the unit sphere. Hence, the parameter space must be $R^d \bigcup \{\infty\}$. It is, however, useful when variational culculus is applied for random fields parametrized by a smooth simple surface.

[2] On the other hand, there are good reasons why κ should not be involved in the algebra $\mathbf{g}$.

i) From our viewpoint that the Fourier transform is particularly emphasized. So the complex Schwartz space is fitting for the complex analysis. Namely, the Schwartz space S_c, which is invariant under the Fourier transform, is more significant. While, in order to introduce the κ we need another space like D_0, instead of S_c.

ii) Under the Fourier transform we have a formal adjoint

$$u^2\frac{d}{du} + u \mapsto \lambda\frac{d^2}{d\lambda^2} + \frac{d}{d\lambda},$$

up to a constant i. Operators of the third order are not favorable for us.

iii) Observe the following commutation relations.

$$[\kappa, \pi] = 2u^2,$$

$$[\kappa, f] = (2u^2 + u)\frac{d^2}{du^2} + 2\frac{d}{du} + (2u^3 - \frac{u}{2}),$$

$$[\kappa, \sigma] = -2u\frac{d^2}{du^2} - 2\frac{d}{du} + u^3.$$

Again, we see that these are not favorable results for us if the algebra is expected to remain within a class of finite dimensional Lie algebras.

Factorization of f and σ.

We have

$$-2(f - I) = (\frac{d}{du} + u)(\frac{d}{du} - u)$$

and

$$2(\sigma - iI) = (\frac{d}{du} + iu)(\frac{d}{du} - iu).$$

These factorizations would have meaning and be useful.

6.4 Applications

There are many applications of the subgroups ($\subset U(E_c)$) involving whiskers and others. For instance,

1) description of invariance of phenomena in quantum dynamics,
2) description of symmetries of typical differential equations,
3) characterizations of stochastic processes and random fields, and

4) analysis on the half plane of E_c^*.

We now briefly mention examples of cases 2) and 4), emphasizing the importance of the infinite dimensional unitary group. For details we refer to the monograph[40] Chapter 7 and the lecture notes[38].

I. Symmetry of the heat equation and the Schrödinger equation.

First, the idea due to W. Miller, Jr. should be referred. To fix the idea we start with the heat equation

$$\frac{\partial}{\partial t}v(x,t) = \frac{1}{2}\frac{\partial^2}{\partial x^2}v(x,t). \tag{6.4.1}$$

Set

$$Q = \frac{\partial}{\partial t} - \frac{1}{2}\frac{\partial^2}{\partial x^2}.$$

Take a differential form L such that

i) $L = X(x,t)\frac{\partial}{\partial x} + T(x,t)\frac{\partial}{\partial t} + U(x,t)$,
ii) X, T, U are analytic in (x,t),
iii) $Qv = 0$ implies $QLv = 0$.

The collection of such L's spans a vector space and the space is denoted by $\mathcal{G}$. The space $\mathcal{G}$ is a Lie albegra under the product $[\cdot,\cdot]$.

Proposition 6.5 *A differential form L of i) belongs to $\mathcal{G}$, if and only if there exists an analytic function $R(x,t)$ such that*

$$[L, Q] = RQ. \tag{6.4.2}$$

Proof is just a rephrasement of the definition. For details see Miller's article in Encyclopaedia Math.

Definition 6.5 A local group G generated by

$$\{e^{t_1 L_1} e^{t_2 L_2} \cdots e^{t_n L_n}; t_j \in R, L_j \in \mathcal{G}\}$$

is called the symmetry group of Q and is denoted by $G = G(Q)$.

According to W. Miller, the Lie algebra of the symmetry group G of the heat equation is spanned by

$$\begin{aligned} &I, \qquad L_{-2} = \frac{\partial}{\partial t}, \quad L_{-1} = \frac{\partial}{\partial x}, \\ &L_0 = x\frac{\partial}{\partial x} + 2t\frac{\partial}{\partial t}, \quad L_1 = t\frac{\partial}{\partial x} + x, \\ &L_2 = tx\frac{\partial}{\partial x} + t^2\frac{\partial}{\partial t} + (x^2+t)/2. \end{aligned}$$

Our white noise approach to the Lie algebra of the symmetry group is as follows. Define the following mapping T:

$$\zeta \longrightarrow T\zeta = v, \quad \zeta \in S_c,$$

where $v = v(x, t; \zeta)$.

Let $\{g_t\}$ be a whisker in $U(S_c)$. Then, $Tg_t\zeta$ will be expressed in the form $v(x, s; g_t\zeta)$. The infinitesimal generator of Tg_t is easily computed. Thus, we have

Theorem 6.2 *Under the mapping T we have the following correspondence*

$$\begin{aligned} &I \mapsto I, \\ &s \mapsto -\frac{\partial}{\partial x}, \\ &\pi \mapsto t\frac{\partial}{\partial x} + x, \\ &\tau \mapsto 2t\frac{\partial}{\partial t} + x\frac{\partial}{\partial x} + \frac{1}{2}, \\ &f \mapsto (t^2-1)\frac{\partial}{\partial t} + tx\frac{\partial}{\partial x} + \frac{1}{2}(x^2+t-1), \\ &\sigma \mapsto (t^2+1)\frac{\partial}{\partial t} + tx\frac{\partial}{\partial x} + \frac{1}{2}(x^2+t-i). \end{aligned}$$

Proof is easy and is omitted.

Remark 6.2 *Some vague interpretastion of the operator σ was given before in connection with the Fourier-Mehler transform, however we now* *understand the significant participation in the symmetry of the heat equation.*

White noise approach to the symmetry of Schrödinger equation can also be discussed in the similar manner. Actual formulas are obtained in the cases of

1) a free particle,

2) a free particle in a constant external field,

3) the harmonic oscillator,

4) a particle in the potential of repulsive force,

and so on.

Theorem 6.3 *The symmetry of a free particle is expressed by the following table.*

$$
\begin{aligned}
I &\mapsto I \\
s &\mapsto L_s = -\frac{\partial}{\partial x} \\
\pi &\mapsto L_\pi = it\frac{\partial}{\partial x} + x \\
\tau &\mapsto L_t = 2t\frac{\partial}{\partial t} + x\frac{\partial}{\partial x} + \frac{1}{2} \\
f &\mapsto L_f = i(t^2+1)\frac{\partial}{\partial t} + itx\frac{\partial}{\partial x} + \frac{1}{2}(x^2 + it - 1) \\
\sigma &\mapsto L_\sigma = i(t^2-1)\frac{\partial}{\partial t} + itx\frac{\partial}{\partial x} + \frac{1}{2}(x^2 + it - i)
\end{aligned}
$$

The symmetry of the Schrödinger equation with potential can be given in the similar manner to the case of a free particle. See Hida[40] Section 7.5.

II. Analysis on half plane of E_c^*.

In Section 6.2. we introduced the space H^h involving holomorphic polynomials in complex white noise z in E_c^*. Take a linear group $SL(2,R)$ and define a mapping of $z \in E_c^*$: For $\alpha \in SL(2,R)$, let it be expressed in the form

$$\begin{pmatrix} a & b \\ c & d \end{pmatrix}.$$

For $z = x + iy$ define $g_\alpha z$ by

$$g_\alpha z = \frac{az+b}{cz+d}.$$

Then, the upper half plane $E_c^{*+} = \{z \in E_c^*;\ y > 0\}$ is carried into itself by g_α.

Now interesting questions arise.

(1) A class of meromorphic functionals may be considered by forming a functional of the form

$$g_\alpha \varphi(z) = \varphi(g_\alpha z),$$

where $\varphi(z)$ is a harmonic polynomial.

(2) The boundary value of holomorphic polynomials on the line $\{z = x + iy; y = 0\}$ can be considered. Then, we can discuss a generalization of the theorem on Hilbert transform.

Chapter 7

Characterization of Poisson noise

7.1 Preliminaries

Poisson process $P(t), t \geq 0$, where $P(t) = P(t, \omega)$, on (Ω, B, P) has been introduced in Chapter 2, Section 2.1. It is an additive process, indeed a Lévy process, such that the increment $\Delta P = P(t+h) - P(t), h > 0$, is subject to a Poisson distribution:

$$P(\Delta P = k) = \frac{(\lambda h)^k}{k!} e^{-\lambda h}, \quad k = 0, 1, 2, \cdots$$

where $\lambda (> 0)$ is the *intensity*.

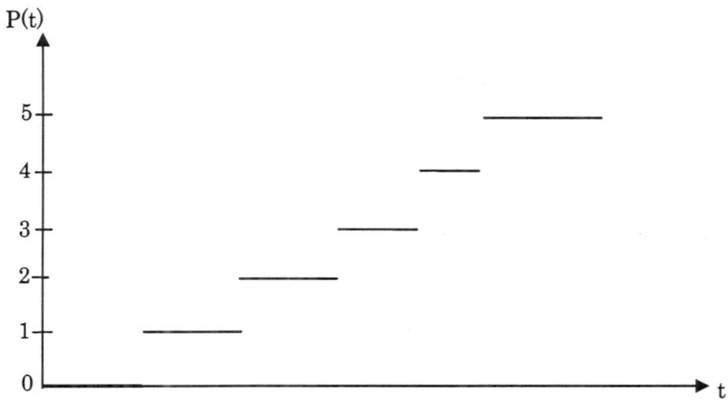

Fig. 7.1.

A sample function of $P(t)$ is a step function such that it starts from 0 and increases only by jumps with unit height. Hence, it is an increasing step function as is shown above in Fig. 7.1. We assume that sample functions are right continuous.

Observing this diagram one feels that it is difficult to imagine latent traits or characteristics of Poisson process. We have, however, a technique, like in the case of (Gaussian) white noise, that makes it a noise, which is obtained by taking time derivative $\dot{P}(t)$. It is called a *Poisson noise.* We shall see that it is easier for us to find its characteristics rather than Poisson process itself, although $\dot{P}(t)$ is a generalized stochastic process. The reasons why noise is easier and the trick to discover hidden characteristics shall be explained step by step, in what follows.

Consider the Poisson noise with the parameter space R^d. Let E_1 be Hilbert space such that it is dense in $L^2(R^d)$ and that there exists an Hilbert-Schmidt injection from E_1 to $L^2(R^d)$ then we have Gel'fand triple

$$E_1 \subset L^2(R^d) \equiv E_0 \subset E_1^*. \tag{7.1.1}$$

The characteristic functional of Poisson noise with R^d parameter is

$$C_P^d(\xi) = \exp\left[\lambda \int_{R^d} (e^{i\xi(t)} - 1) dt^d\right]. \tag{7.1.2}$$

We can prove that $C_P^d(\xi)$ is continuous on E_0. Obviously it is positive definite and $C_P^d(0) = 1$. Thus, the Bochner-Minlos theorem tells us that, "there exists a probability measure μ_P which is supported by E_1^*", hence we have a *Poisson noise* (measure space) (E_1^*, μ_P^d) and μ_P^d is called *Poisson measure*.

We shall take a time parameter space to be a compact set, say $[0, 1]$ to fix the idea in what follows. The Gel'fand triple can therefore be replaced by

$$E_1 \subset L^2([0,1]) \equiv E_0 \subset E_1^*. \tag{7.1.3}$$

The definition of nuclear spaces etc. can be seen in Appendix I.

The Poisson noise measure μ_P^1 will be simply denoted by μ_P.

Now it is easy to see a direct connection between Poisson noise (E_1^*, μ_P') and Poisson process $P(t)$ as follows.

The sample function $x \in E_1^*$ of a Poisson noise $\dot{P}(t)$ is obtained from a Poisson process $P(t)$, over a unit time interval $[0, 1]$. The stochastic bilinear form $P(t) = \langle x, \chi_{[0,t)} \rangle$ is a version of a Poisson process $P(t), 0 \le t \le 1$. If we come to higher dimensional parameter case like R^d, such a connection is not as is expected to define a Poisson field, but a Poisson sheet, which is an integral of a Poisson noise.

Poisson noise might be able to give us any rich group that describes its invariance compared to (Gaussian) white noise. One can only see the shift invariance other than pathological transformations if we consider a Poisson noise over the time interval $(-\infty, \infty)$. There is, however, a different viewpoint to see the invariance; in fact this is more essential.

As in the paper[148], the *partition* of the sample space E^* is the key technique to find invariance. Put the shift operator aside, and let the parameter set be restricted to a finite interval, say [0,1]. Then we can count the number of jumps, which is finite, and the periods of jump, for a sample function of $P(t, x)$. With these statistics we can characterize a Poisson process. We give a quick interpretation of the idea of characterization. More details will be discussed in the next section.

Consider an event A_n defined by

$$A_n = \{x \in E^*; P(1, x) = n\}. \tag{7.1.4}$$

A conditional probability space $(A_n, \mathbf{B}_n, P_n)$ is given, where

$$\mathbf{B}_n = \{B \bigcap A_n; B \in \mathbf{B}\}$$

and $\mu_P^n(B) = \mu_P(B|A_n)$. On the conditional probability space it can be said that the n points of discontinuities (or jump points) of $P(t)$ are equally distributed. Namely, the joint distribution is invariant under the symmetric group $S(n)$ acting on the coordinates of those points.

Therefore, the total invariance for $P(t)$ may be expressed by the product group G:

$$G = \bigotimes_{n=0}^{\infty} S(n) \tag{7.1.5}$$

acting on the product set $\prod_{n=0}^{\infty} A_n$, where $S(0)$ is understood to be a trivial group involving only identity.

Note that the event A_n is viewed as an event where $\dot{P}(t)$ consists of delta functions of exactly as many as n. The actions of the group $S(n)$ are permutations of n delta functions.

We further think of the unitary representation of the product group of symmetric groups. It is our hope that the representation would contribute to the analysis of Poisson noise functionals.

In view of such a setup, it is easy to discuss the analysis of functionals of Poisson noise where the significant results by Y. Ito and I. Kubo play a dominant role. It is noted that we can play the same game in the case of R^d-parameter Poisson noise.

7.2 A characteristic of Poisson noise

Our purpose is to characterize a Poison noise. Note the characterization of white noise, using rotation group $O(E)$, which is well known, see e.g. Hida[40]. There should be significant difference between the cases of Gaussian and Poisson type in probabilistic character.

To concretize the idea explained above, we will try to find a suitable transformation group which is acting on the Poisson noise measure space and possibly characterizes Poisson noise measure itself. A rigorous setup of our survey is as follows.

Let A_n be the event on which there are n jump points over a finite time interval $I = [0, 1]$. It is defined by (7.1.4), where n is any non-negative integer.

Then, the collection of the events $\{A_n, n \geq 0\}$ is a partition of the entire sample space E_1^*. Namely, up to μ^P-measure 0, the following relationships hold:

$$A_n \bigcap A_m = \phi, \ n \neq m; \ \bigcup A_n = E_1^*. \tag{7.2.1}$$

The conditional probability $\mu_P^n(A)$ given A_n is

$$\mu_P^n(A) = \frac{\mu_P(A_n \cap A)}{\mu_P(A_n)}. \tag{7.2.2}$$

If $C \subset A_k$, the probability measure is

$$\mu_P^k(C) = \mu_P(C|A_k) = \frac{\mu_P(C)}{\mu_P(A_k)}$$

on a probability measure space $(A_k, \mathbf{B}_k, \mu_P^k)$, where $\mathbf{B}_k$ is a sigma field generated by measurable subsets of A_k, determined by $P(1, x)$.

For $k = 0$, the measure space is trivial. It is $(A_0, \mathbf{B}_0, \mu_P^0)$ and

$$\mathbf{B}_0 = \{\phi, A_0\} \bmod \mu_P^0$$

with $\mu_P^0(A) = 1$.

Now we give a general concept concerning probability measure space.

Definition 7.1 A probability space $(\Omega, \mathbf{B}, P)$ is a *Lebesgue space* if

1) There is a countable base $\mathbf{I} = \{I_n\}$ such that $\mathbf{B} = \sigma\{I_n; n \in Z\}$, where $\sigma\{\ \}$ means the sigma field generated by the sets in $\{\cdot\}$.
2) If $A_n = I_n$ or I_n^c, then $\cap_n A_n \neq \emptyset$.
3) For $x \neq y$, there exist I_n, I_m such that $I_n \cap I_m = \emptyset$, and that $x \in I_n$, $y \in I_m$.

Example 7.1 The system of sets defining the Rademacher functions forms a base of the Lebesgue measure space $([0,1], \mathbf{B}, dm)$, m being the Lebesgue measure.

The following theorem is well known.

Theorem 7.1 *Let $(\Omega, \mathbf{B}, P)$ be a Lebesgue space. Then it is isomorphic to $([0,a], Leb) \cup (atoms)$ for some $a \geq 0$, where atom means a measurable set that has positive measure and has no $\mathbf{B}$-measurable proper subset.*

Returning to the partition $\{A_k\}$ determined by the Poisson process $P(t,x)$, we claim

Proposition 7.1 $(A_k, \mathbf{B}_k, \mu_P^k)$, *$k \geq 1$ is a Lebesgue space without atom.*

Proof comes from the fact that $(A_k, \mathbf{B}_k, \mu_P^k)$ is isomorphic to the n-times direct product of uniform measure on $[0,1]$.

Remark 7.1 *On a Lebesgue space, roughly speaking, a measure preserving set transformation has a version of a measure preserving point transformation.*

Let x be in $A_n, (n \geq 1)$, and let $\tau_i, 1 \leq i \leq n$, be the order statistics of jump points of $P(t, x)$:

$$\tau_i \equiv \tau_i(x), i = 1, 2, ..., n.$$

We therefore have the inequalities

$$0 = \tau_0 < \tau_1 < \cdots < \tau_n < \tau_{n+1} = 1$$

almost surely. Set

$$X_i(x) = \tau_i - \tau_{i-1}, \ 1 \leq i \leq n+1,$$

so that

$$\sum_1^{n+1} X_i = 1.$$

Note that $P(t, x)$ is temporary homogeneous additive processes. Hence the first jump point τ is uniformly distributed on [0,1]. The conditional probability distribution given τ_1 is uniform on $[\tau_1, 1]$ and so forth.

We can now easily prove

Proposition 7.2 *Under the assumption $P(1) = n$, that is on the probability space $(A_n, \mathbf{B}_n, \mu_P^n)$, the probability distribution of the random vector $(X_1, X_2, ..., X_{n+1})$ is uniform on the simplex*

$$\sum_{j=1}^{n+1} x_j = 1, x_j \geq 0.$$

Corollary 7.1 *The probability distribution function of each X_j is*

$$1 - (1-u)^n, 0 \leq u \leq 1.$$

(See the probability distributions of order statistics $\tau_j, 1 \leq j \leq n+1$.)

Proposition 7.3 *The conditional characteristic functional of $\dot{P}(t)$ given A_n*

$$C_{P,n}(\xi) = E\left(e^{i\langle \dot{P}, \xi\rangle} | A_n\right), \quad \xi \in E_1, \tag{7.2.3}$$

is expressed in the form

$$C_{P,n}(\xi) = \left(\int_0^1 e^{i\xi(t)} dt \right)^n. \tag{7.2.4}$$

Proof. Each jump point of $P(t)$ is distributed uniformly on $[0, 1]$, and those jumps are independent. This implies the required expression.

■

Remark 7.2 *The conditional distribution given A_n is intensity-free for every n.*

Transformations that keep the Poisson measure invariant

We try to find a transformation group that is acting on the Poisson noise space, (E^*, μ_P), keeping the μ_P invariant as was given in the last section.

Note that in Gaussian case the transformation g^* acts on each sample functions $x \in E^*$, of $\dot{B}(t)$, in such a way that

$$g^* : x \to g^* x, \ g^* \in O^*(E^*).$$

For the case of a Poisson noise, in contrast to Gaussian case, we begin with measure preserving set transformations and eventually, we can come to the transformations of sample functions, by using the Lebesgue space structure.

Our plan is that we first investigate the invariance of Poisson noise measure, and then we aim at harmonic analysis of functionals of Poisson noise, where the group of μ_P-invariant transformations will play significant roles. In particular, the family of symmetric groups acting on the conditional probability spaces describes a characteristic of Poisson noise.

There are two ways of introducing such a transformation group. The idea is the same as in the rotation group $O(E)$, i.e. subgroups are classified into two classes as in Section 5.6. One class is defined by using the vector space structure of E^* with the help of a particular base of E^*. Another class comes from the transformations acting on the parameter space. This means that the second method, which is unlike the first method, depends on the geometric structure of the space in question.

Both methods present the characteristics of Poisson noise, however their probabilistic appearances are quite different from Gaussian case. Thus, the groups describing the invariance of Poisson noise will naturally appear in two ways : Class I and Class II.

To make the discussion somewhat general, we take the parameter space to be multi-dimensional Euclidean space, say R^d.

Class I. Use of an orthonormal system

We give an illustrative example that is defined by using an orthonormal system of $L^2(R^d)$. Difference from case II below is worth to be noted.

Let $\{\xi_n(t)\}, n = 1, 2, \cdots$, be a particular orthonormal (not complete) system in $L^2(R^d)$ defined in the following manner. Take domains $I_\alpha = \Pi_{j=1}^d [k_j, k_{j+1}) \subset R^d, \alpha = (k_1, k_2, \cdots, k_d),\ k_j \in Z$. Note that $\{I_\alpha\}$ forms disjoint domains in R^d. Suppose that I_α's are ordered linearly and they are enumerated as $I_n, n \geq 1$. Set

$$\xi_n(t) = \chi_{I_n}(t). \tag{7.2.5}$$

Let π be a permutation of finitely many n and g_π be the transformation such that

$$(g_\pi \xi)(t) = \sum_{-\infty}^{\infty} a_n(t) \xi_{\pi(n)}(t) \in L^2(R^n), \tag{7.2.6}$$

where

$$\xi(t) = \sum_{-\infty}^{\infty} a_n(t) \xi_n(t) \ \in \ L^2(R^d).$$

A stochastic bilinear form $\langle x, \xi_n \rangle$ defines a Poisson random variable denoted by $X_n(x)$. Similarly, we can define $X_{\pi(n)}(x)$ that is subject to the same Poisson distribution. Let $\mathbf{B}(X_n)$ be the sigma-fields with respect to which X_n is measurable. The g_π defines a measure preserving transformation from $\bigvee_n \mathbf{B}(X_n)$ to itself, since $\bigvee_n \mathbf{B}(X_n)$ is invariant under the action of g_π. Thus, we have defined an invariance g_π of μ_P.

Although g_π is not a continuous mapping on E, it defines a μ_P-measure preserving transformation, let it be denoted by g^* symbolically.

Note that there are many ways of generalization of g_π with this idea, and they illustrate some invariance of Poisson noise. In any case, invariance is guaranteed in the weak sense, i.e. in law.

Class II. Group of motions

We can easily see that the time shift leaves the probability distribution μ_p of a Poisson noise invariant, since

$$C_p(S_t^k\xi) = C_p(\xi), \quad (S_t\xi)(u) = \xi(u - te_k),\ u \in R^d, \tag{7.2.7}$$

where e_k is the unit vector in the k-th direction of R^d.

Obviously, orthogonal group $O(d)$ acting on the parameter space R^d presents an invariance of Poisson measure, since the characteristic functional $C_P^d(\xi)$ is kept invariant.

Thus, we have

Proposition 7.4 *The probability distribution μ_P of an R^d-parameter Poisson noise is invariant under*
1) the orthogonal group $O(d)$, and
2) the shifts of R^d.

Note. Interesting property can be seen in the isotropic dilation $\tilde{\tau}_t$ of R^d-vectors:

$$\tilde{\tau}_t : u \to ue^{at},\ u \in R^d,\ a > 0. \tag{7.2.8}$$

Then characteristic functional $C_P^d(\xi)$ changes to

$$\exp[\lambda \int (e^{\xi(ue^{at})} - 1)du^d] = \exp[\lambda e^{-dat} \int (e^{\xi(v)} - 1)dv^d]. \tag{7.2.9}$$

Namely, the intensity changes from λ to λe^{-dat}, but the measure remains to be a Poisson distribution.

T^d-parameter case

Take the parameter space to be a compact domain with a suitable geometric structure, say T^d, d-dimensional Torus instead of R^d in (7.1.1). Then we have a corresponding Gel'fand triple

$$E_1 \subset L^2(T^d) \equiv E_0 \subset E_1^*. \tag{7.2.10}$$

The characteristic functional of Poisson noise with T^d parameter is

$$C_P(\xi) = \exp\left[\lambda \int_{T^d} (e^{i\xi(t)} - 1)dt\right].$$

Example 7.2 Take, in particular, $d = 1$. Then, T^d is S^1, a circle, and the characteristic functional is

$$C_{P,1}(\xi) = \exp\left[\int_0^{2\pi} (e^{i\xi(t)} - 1)dt\right],$$

where intensity λ is taken to be 1, to be specified.

Let τ be a transformation defined by

$$(\tau)(t) = pt(\text{mod } 2\pi), t \in S^1, \tag{7.2.11}$$

where p is a positive integer. Define a mapping σ such that

$$(\sigma\xi)(u) = \xi(\tau(u)). \tag{7.2.12}$$

Then we can see that

$$C_{P,1}(\sigma\xi) = C_{P,1}(\xi).$$

This means that the characteristic functional is σ-invariant. Since σ is continuous on E_1, the σ^* is defined, and hence the measure μ_P is σ^*-invariant.

We have some more general assertions.

Proposition 7.5 *The mapping σ is a continuous automorphism of E_1 such that $F(\sigma\xi) = F(\xi)$, where*

$$F(\xi) = \int_{T^d} (e^{i\xi(t)} - 1)dt,$$

then there exists a measurable measure σ^* on E^* such that

$$\sigma^* \mu_P = \mu_P.$$

Poisson measure preserving transformations reduced to A_n

We restrict our attention to the event $A_n, n \geq 1$, which is a member of the partition of E^* introduced in (7.1.4).

Let

$$S(n+1) = \{\pi^n(i), 1 \leq i \leq n+1\}$$

be the symmetric group, where $\pi^n(i)$ is a permutation of integers $1, 2, \cdots, n+1$, and let $\mathbf{B}_n = \mathbf{B}(\Delta_{n+1})$ be the sigma field of all Borel subsets of simplex types

$$\Delta_n = \left\{(x_1, x_2, \cdots, x_{n+1} \mid \sum_1^{n+1} x_i = 1,\ 0 \leq x_i \leq 1\right\}.$$

Let

$$X_1(x), X_2(x), \cdots, X_{n+1}(x), \quad x \in A_n,$$

be the time intervals ($\subset [0,1]$) divided by the successive n instants of jump of the Poisson process $P(t,x)$, as in Proposition 7.1. Set

$$\mathbf{X} = \mathbf{X}(x) = (X_1(x), X_2(x), \cdots, X_{n+1}(x)).$$

Then, for a permutation π^n in $S(n+1)$, such that

$$j \rightarrow \pi_j^n$$

we can define a mapping

$$\mathbf{X} \mapsto \mathbf{X}_{\pi^n}(x) = \left(X_{\pi_1^n}(x), X_{\pi_2^n}(x), ..., X_{\pi_{n+1}^n(x)}\right),$$

which determines a set transformation g_{π^n} on E^* in such a way that

$$g_{\pi^n} : \mathbf{X}^{-1}(B) \mapsto \mathbf{X}_{\pi^n}^{-1}(B), \quad B \in \mathbf{B}_n.$$

Note that $\mathbf{X}^{-1}(B)$ and $\mathbf{X}_{\pi^n}^{-1}(B)$ are measurable subsets of A_n.

By Proposition 7.2 we can see that

$$\mu_P^n(X^{-1}(B)) = \mu_P^n(X_{\pi^n}^{-1}(B)),$$

holds and g_{π^n} defines an automorphism of (A_n, μ_P^n).

Then an automorphism g_π on E^* is defined with the help of the automorphism g_{π^n} on $A_n, n = 1, 2, \cdots$, such that

$$g_\pi x = g_{\pi^n} x, \quad x \in A_n, \quad n = 0, 1, 2, ...$$

Since the space $(A_k, \mathbf{B}_k, \mu_P^k)$ is a Lebesgue space, the g_{π^k} defines a measure preserving point transformation on $(A_k, \mathbf{B}_k, \mu_P^k)$. The same for g_π acting on E_1^*.

Then, we have

Proposition 7.6 *The transformation g_π is a μ_P-measure preserving transformation on E_1^*.*

Set

$$\begin{aligned} \mathbf{G}_k &= \{g_{\pi^k}; \pi^k \in S(k+1)\} \\ \mathbf{G} &= \{G_1, G_2, \cdots, G_k, \cdots\} \end{aligned} \tag{7.2.13}$$

then

$$\begin{aligned} \mathbf{G}_k &\cong S(k+1), \\ \mathbf{G} &\cong \{S(2), S(3), \cdots, S(k+1), \cdots\}. \end{aligned} \tag{7.2.14}$$

Summing up, we have

Theorem 7.2 *The group* **G** *defines a symmetry of Poisson noise.*

The group **G** may be regarded as the same as the *infinite symmetric group* which is defined as an inductive limit of finite symmetric group with the fixed structure. The group actually gives an invariance of Poisson noise measure. Another way to understand an infinite symmetric group is discussed in connection with *duality* between Gaussian and Poisson noises. Details of this topic has appeared in the paper[149]. In those cases we can see an aspect of characterization of Poisson noise.

7.3 A characterization of Poisson noise

We shall give a characterization of Poisson noise by using the transformation like Baker's transformation.

For a while, the basic probability space is not specified to be (E_1^*, μ_P), but a general space with probability parameter $\omega \in \Omega$.

Suppose we are given a partition $\{G_n, n \geq 0\}$ of Ω with $P(G_n) > 0$ for every n. The value of $P(G_n)$ will be determined later. Start with conditional probability space $(G_n, P_n), n \geq 0$, where a stochastic processes $\{Y_n(t,\omega), \omega \in G_n\}$ is defined, in the following manner:

(1) $Y_n(t,\omega) = \sum_1^n \delta_{t_j(\omega)}(t),\ t \in [0,1]$,
(2) the joint probability distribution of n singular points $t_j(\omega), j = 1, \cdots, n$, are absolutely continuous,
(3) $\{Y_n(t,\omega)\}$ and $\{Y_n(\tau'(t),\omega)\}$ have the same probability distribution, where τ' is the transformation defined by

$$\tau'(t) = 2t(\text{mod}1). \tag{7.3.1}$$

Theorem 7.3 *Under the above assumptions (1) $\sim$ (3), the probability distribution of each singular point $t_j(\omega), \omega \in G_n$ is a uniform distribution on $[0,1]$ and t_j's are invariant.*

Proof. For $n = 0$, it is trivial.
For $n = 1$, only one delta function is involved. Let $p(t)$ be the probability distribution of $t_1(\omega)$. Use the transformation τ', defined by (7.3.1), for which p is invariant, so that we have the iterative formula

$$p(t) = \frac{1}{2}\left(p(\frac{t}{2}) + p(\frac{t}{2} + \frac{1}{2})\right). \tag{7.3.2}$$

After N iterations of the above formula, it is obtained as

$$p(t) = \frac{1}{2^N}\sum_{k=0}^{2^N-1} p(\frac{t+k}{2^N}). \tag{7.3.3}$$

Let N tend to infinity then we have

$$p(t) = \int_0^1 p(u)du,$$

by using the assumption (2). This result means that $p(t)$ is a constant function, and consequently it is a uniform distribution.

This result can be generalized to the case for general n. The probability density function $p(t_1, t_2, \cdots, t_n)$ of $(t_1, t_2, \cdots, t_n)$ satisfies the equation

$$p(t_1, t_2, \cdots, t_n) = \sum_{p=1}^{n}\frac{1}{2}\left(p(t_1, \cdots, \frac{t_p}{2}, \cdots, t_n) + p(t_1, \cdots, \frac{t_p}{2} + \frac{1}{2}, \cdots, t_n)\right). \tag{7.3.4}$$

Hence we have

$$p(t_1, t_2, \cdots, t_n) = \frac{1}{2^{Nn}} \sum_{p=1}^{n} \sum_{k_p=0}^{2^N-1} p(\frac{t_1+k_1}{2^N}, \frac{t_2+k_2}{2^N}, \cdots, \frac{t_p+k_p}{2^N}, \cdots, \frac{t_n+k_n}{2^N}). \tag{7.3.5}$$

Letting $N \to \infty$, we have

$$p(t_1, t_2, \cdots, t_n) = \int_0^1 \cdots \int_0^1 p(u_1, u_2, \cdots, t_n))du^n,$$

which means $p(t_1, t_2, \cdots, t_n)$ is a constant function. Thus the assertion has been proved.

■

Proposition 7.7 *The characteristic functional $C_n(\xi)$ is invariant under the transformation σ', where σ' is defined by*

$$(\sigma'\xi)(u) = \xi(\tau'(u)).$$

Using independence and uniform distribution of the random variables t_j's, we can prove

Theorem 7.4 *The conditional characteristic functional of $Y_n(t)$ on the space G_n with the parameter space (time interval) [0,1] is*

$$C_n(\xi) = \left(\int_0^1 e^{i\xi(u)} du\right)^n. \tag{7.3.6}$$

We now take the parameter space $[0, t]$ instead of $[0, 1)$.

Corollary 7.2 *The characteristic functional of $Y_n(t)$ on the space G_n is expressed in the form*

$$C_n(\xi, t) = \left(\frac{1}{t} \int_0^t e^{i\xi(u)du}\right)^n. \tag{7.3.7}$$

We recall that $\Omega = \sum_n G_n$. Let us define a stochastic process $Y(t)$ on the entire space (Ω, P) such that

(i) $Y(t, \omega) \equiv Y_n(t, \omega), \ \omega \in (G_n, P_n)$.

(ii) $Y(t)$ is additive and temporary homogeneous.

In order to have $Y(t)$ well defined, it is necessary to determine the probability of G_n.

Let $P(G_n) = c_n$. The characteristic functional $C(\xi)$ on Ω is given by

$$C(\xi, t) = \sum_n c_n C_n(\xi, t), \tag{7.3.8}$$

since $C_n(\xi, t) = E(e^{i\xi(t)}|G_n)$. ¿From the above assumption (ii), that is $Y(t)$ is additive and temporary homogeneous we have

$$C(\xi, s+t) = C(\xi, t)C(\xi, s). \tag{7.3.9}$$

The relation (7.3.6) and (7.3.9) yields

$$\sum_n c_n \left(\frac{1}{s+t}\int_0^{s+t} e^{i\xi(t)}\right)^n$$
$$= \left(\sum_j c_j \left(\frac{1}{s}\int_0^s e^{i\xi(t)}\right)^j\right)\left(\sum_k c_k \left(\frac{1}{t}\int_0^t e^{i\xi(t)}\right)^k\right).$$

Because of the temporary homogeneity of $Y(t)$, we have

$$\sum_n \sum_l d_n(s+t)\binom{n}{l} a^l b^{n-l} = \sum_j \sum_k d_j(s)d_k(t)a^j b^k$$

where

$$a = \int_0^s e^{i\xi(t)},\ b = \int_0^t e^{i\xi(t)},\ d_k(t) = \frac{c_k}{t^k}.$$

Equating the coefficient of $a^l b^{n-l}$

$$n!d_n(s+t) = (l!d_l(s))((n-l)!d_{n-l}(t)).$$

By letting $h_k(t) = k!d_k(t)$, we obtain that $h_0(t) = 1, h_k(t) = \lambda^k, k = 1, 2, \cdots$, where λ is a constant, Then we are given

$$P(G_k) = c_k = \frac{(\lambda t)^k}{k!}. \tag{7.3.10}$$

It means that $Y(t)$ has a Poisson distribution with

$$P(\cdot|G_n) = P_n(\cdot).$$

Theorem 7.5 *Under the above assumptions, the $Y(t)$ is a Poisson noise.*

Proof. From (7.3.6) and (7.3.9) the characteristic functional of $Y(t)$ is obtained as

$$C(\xi) = \exp[\lambda \int_0^t (e^{i\xi(u)} - 1)du],$$

which proves the theorem. ∎

Reproducing kernel Hilbert space method

The time parameter space is kept to be $[0, 1]$. The characteristic functional of Poisson noise has a Taylor series expansion as follows:

$$C^P(\xi) = \sum_0^\infty \frac{\lambda^n}{n!} e^{-\lambda} C_n(\xi), \tag{7.3.11}$$

where $C_n(\xi) = (\int_0^1 e^{i\xi(u)}du)^n$.

The functionals $C^P(\xi)$ and $C_n(\xi), n = 0, 1, 2, \cdots$, are all positive definite, so they define reproducing kernel Hilbert spaces (RKHS), denoted by $\mathcal{F}$ and $\mathcal{F}_n$, respectively.

$$\mathcal{F} = S\left\{f(\cdot) = C^P(\cdot - \xi);\ \xi \in E\right\},$$

where S means the space spanned by the members in the curly bracket. The topology of the space is determined by the Hilbertian norm that comes from the inner product $(f(\cdot), C^P(\cdot - \xi))_{\mathcal{F}} = f(\xi)$.

In a similar manner we can form the RKHS $\mathcal{F}_n$ by using the kernel $C_n(\cdot - \xi)$.

Proposition 7.8 *The RKHS $\mathcal{F}_n$ is a subspace of $\mathcal{F}$.*

Proof. Although the C^P has real structure as a functional on E, we can take $\xi + i\log a$ (a being a positive constant) as a variable of C^P, since the time interval is $[0, 1]$. Hence, $\frac{d^n}{da^n}C^P(\cdot - (\xi + \log(ia)))$ tends to $\lambda^n e^{-\lambda t} C_n(\cdot - \xi)$ as $a \to 0$. Hence, the generating member of $\mathcal{F}_n$ is in $\mathcal{F}$. This implies the assertion. ∎

Proposition 7.9 *i) The subspaces $\mathcal{F}_n$'s are mutually orthogonal.*

ii) $\mathcal{F} = \oplus \mathcal{F}_n$.

Proof is easy, so is omitted.

In terms of characterization, we may say that the characteristic functional is determined by the Taylor series expansion in terms of the powers $(\int e^{i\xi(u)} du)^n, n \geq 0$.

Remark 7.3 *The characteristic functional that appeared in Theorem 7.5 which generates the subspace $\mathcal{F}_n$ is an eigenfunctional of the Lévy Laplacian Δ_L. (Refer to the literature[134].) Thus, we may simply say that Poisson noise is characterized by the eigenfunctionals of the Lévy Laplacian with eigenvalues $-n$. Note that Lévy Laplacian is derived eventually from (Gaussian) white noise.*

7.4 Comparison of two noises; Gaussian and Poisson

Some properties reviewed so far in this chapter are taken to be characteristics of Poisson noise. Among others, uniform probability distribution on a simplex and the invariance of the conditional Poisson distribution under the symmetric group are significant to characterize a Poisson noise.

With this fact in mind, we are going to construct a Poisson noise in such a way that we decompose a measure space; that is, from a partition of Ω defining the noise and on each component of the partition we can see the characteristics in a visualized manner.

Also our method can be compared to the approximation to Gaussian white noise by projective limit of spheres, where one can see the characteristics; namely, rotation invariant and maximum entropy. The comparison is successful to observe the two noises step by step in the course of approximation. There latent properties reveal in front of us.

We first observe the Gaussian case by reviewing the result by Hida-Nomoto [63].

G.0) Start with a measure space $(S_1, d\theta)$, where S_1 is a circle and $d\theta = \frac{1}{2\pi}$.
G.1) Take a measure space $(\tilde{S}_n, \sigma_n)$, where $\tilde{S}_n$ is the n-dimensional space

excluding north and south poles; and where $d\sigma_n$ is the uniform probability measure on $\tilde{S}_n$.

G.2) Define the projection $\pi_n : \tilde{S}_{n+1} \to \tilde{S}_n$ in such a way that each longitude is projected to the intersection point of the longitude and the equator. Thus, $\tilde{S}_n$ is identified with the equator of $\tilde{S}_{n+1}$. Take the radius of $\tilde{S}_{n+1}$ be $\sqrt{n}$.

G.3) Given the measure σ_n on $\tilde{S}_n$ which is the uniform probability measure, i.e. invariant under the rotation group $SO(n+1)$. Then the measure σ_{n+1} is defined in such a way that

$$\sigma_{n+1}(\pi_n^{-1}(E)) = \sigma_n(E), E \subset \tilde{S}_n$$

and that σ_{n+1} gives the maximum entropy under the above restriction.

G.4) As a result, σ_{n+1} is proved to be the uniform distribution on $\tilde{S}_{n+1}$ and is invariant under the rotation group $SO(n+2)$.

G.5) The projective limit of the measure space is defined by the projections π_n, and actually a measure space $(\tilde{S}_\infty, \sigma)$ is defined

$$(\tilde{S}_\infty, \sigma) = \text{proj}\cdot\lim\ (\tilde{S}_n, \sigma),$$

where "proj$\cdot$lim" denotes the projective limit.

G.6) The limit can be identified with the white noise measure space (E^*, μ) with the characteristic functional $e^{-\frac{1}{2}\|\xi\|^2}$, where E^* is the space of generalized functions.

In parallel with Gaussian case, we can form a Poisson noise by an approximation where optimality (maximum entropy) and invariance under the symmetric group are effectively used.

P.0) Start with a probability space (Δ_1, μ_1), where $\Delta_1 = \{(x_1, x_2); x_i \geq 0,\ x_1 + x_2 = 1\}$ and μ_1 is the Lebesgue measure.

P.1) Define a probability space (Δ_n, μ_n), where Δ_n is a Euclidean n-simplex,

$$\Delta_n = \left\{(x_1, \cdots, x_{n+1}),\ x_i \geq 0,\ \sum_1^{n+1} x_i = 1\right\}$$

in R^{n+1}.

P.2) Define π_n to be the projection of Δ_{n+1} down to Δ_n which is a side simplex of Δ_{n+1}, determined as follows. Given a side simplex Δ_n of Δ_{n+1}. Then, there is a vertex of Δ_{n+1} which is outside of Δ_n, let it be denoted by v_n. The projection π_n is a mapping defined by

$$\pi_n : \overline{v_n x} \to x.$$

where $\overline{v_n x}$ is defined as a join connecting the v_n and a point x in Δ_n.

P.3) We introduce a probability measure μ_n under the requirements that

$$\mu_{n+1}(\pi_n^{-1}(B)) = \mu_n(B), \quad B \subset \Delta_n$$

and that μ_{n+1} has maximum entropy under the boundary condition.

P.4) Since there is a freedom to choose a boundary of Δ_{n+1}, the requirements on μ_{n+1} in P.3), the measure space $(\Delta_{n+1}, \mu_{n+1})$ should be invariant under symmetric group $S(n+2)$ which acts on permutations of coordinates x_i.

P.5) We can form $\{(\Delta_n, \mu_n)\}$ successively by using the projection $\{\pi_n\}$, where μ_n is the uniform probability measure on Δ_n.

Set

$$\Delta_\infty = \cup_n \Delta_n,$$

$$\mathbf{B}(\Delta_\infty) = \sigma\text{-field generated by} \cup A_n, \ A_n \in \mathbf{B}(\Delta_n),$$

and

$$\mu(A) = \sum_n p_n \mu_n(A \cap \Delta_n), \quad A \in \mathbf{B}(\Delta_\infty),$$

where $p_n = \frac{\lambda^n}{n!}$ which is obtained in (7.3.10).

Theorem 7.6

i) The measure space (Δ_n, μ_n) is isomorphic to the measure space (A_n, P_n) defined in Section 7.1.

ii) The weighted sum (Δ_∞, μ) of measure spaces $(\Delta_n, \mu_n), n = 1, 2, \cdots$ is identified with the Poisson noise space.

Proof. According to the facts P.1 to P.6, the theorem is proved. ■

What have been investigated are explained as follows. Basic noises, that is Gaussian and Poisson noises, have

1) optimality in randomness, which is expressed in terms of *entropy*, and
2) invariance under the transformation group; each noise has its own characteristic *group* (one is continuous and the other is discrete).

Such a principle can be discovered in the course of constructing finite dimensional approximation and partition for the two noises in question, respectively. Finally, a certain duality could be found between the two (Professor Ojima's comment).

Remark 7.4 *So far the parameter space is taken to be [0,1]. The characteristic properties that have been discussed can be generalized to the case of R^d-parameter with modest modifications.*

7.5 Poisson noise functionals

We have so far discussed Poisson noise with special emphasis on its characterization. With this background we are now in a position to study the analysis of functionals of Poisson noise.

Much of the result, at least formal approach to calculus, is similar to that of Gaussian case. We shall, therefore focus our attention on dissimilarity between Gaussian and Poisson.

To fix the idea, we shall, in what follows, take the time parameter set to be $I = [0,1]$. Let μ^p be the Poisson measure determined by the characteristic functional

$$C_p(\xi) = \exp\left[\int (e^{i\xi(t)-1)dt}\right], \xi \in E. \tag{7.5.1}$$

We specify (7.1.2) by replacing R^d, dt^d and λ to be I, dt and 1, respectively. The μ^p-almost all x in E^*, is now viewed as sample functions of $\dot{P}(t)$.

The complex Hilbert space $(L_P^2) = L^2(E_1^*, \mu^P)$ is defined in the usual manner. There are many ways to have the Fock space arising from Poisson noise. For instance, it is easy to have a direct sum of decomposition of (L_P^2);

$$(L_P^2) = \bigoplus_{N=0}^{\infty} H_n$$

by using orthogonal polynomials in $\dot{P}(t)$'s. The Charlier polynomials play the same role as Hermite polynomials did in Gaussian case (see e.g. Hida[36].) We shall, in this section, use Y. Ito–I.Kubo's S-transform, denoted by S_P, and called S_P-transform, introduced in the paper[78]. It is defined by

$$(S_P\varphi)(\xi) = \int_{E_1^*} \exp\left[\langle x, \log(1+\xi)\rangle - \int_I \xi(u)du\right] \varphi(x) d\mu^P(x) \qquad (7.5.2)$$

for $\varphi \in (L_P^2)$.

Note that the functional $\exp[\cdots]$ in the above integrand is not a test functional and may not be real, however it can be proved that the integral (7.5.2) is well defined.

The image of (L_P) under S_P is a reproducing kernel Hilbert space (RKHS) denoted by $\mathcal{F}_P$. This fact can be proved as is done in Chapter 2, where we are given the RKHS $\mathcal{F} \sim (L^2)$ (see (2.5.3)).

There is a theorem, namely we have

Theorem 7.7 *(Y. Ito and I. Kubo) The two RKHS's $\mathcal{F}$ and $\mathcal{F}_P$ are the same, and hence S-transform and S_P-transform give an isomorphism*

$$(L^2) \cong (L_P^2). \qquad (7.5.3)$$

By this theorem, we can define generalized Poisson noise functionals to have

$$(L_P^2)^- \cong (L^2)^-. \qquad (7.5.4)$$

In particular, $\dot{P}(t)$ is a member of $(L_P^2)^-$ and no more a formal random variable. We can give it an identity, so that it is a generalized linear functional of Poisson noise. Polynomials in $\dot{P}(t)$'s are also generalized Poisson noise functionals.

Then, we can come to partial differential operators, the formal expression of which are of the form

$$\partial_t = \frac{\partial}{\partial \dot{P}(t)}, \ t \in I.$$

It is defined by

$$\partial_t = S_P^{-1} \frac{\delta}{\delta \xi(t)} S_P,$$

where $\frac{\delta}{\delta \xi(t)}$ is the Fréchet derivative acting on $\mathcal{F}_P$. The domain of ∂_t is rich enough in $(L_P^2)^-$, and it acts as an annihilation operator. Its adjoint, denoted by ∂_t^*, is a creation operator that serves as a stochastic integration operator with respect to Poisson noise.

The multiplication by $x(t)$ or by $\dot{P}(t)$ is expressed in the form

$$x(t) \cdot \; (= \dot{P}(t) \cdot \;) = (\partial_t^* + 1)(\partial_t + 1), \tag{7.5.5}$$

which is somewhat different from Gaussian case (cf. Section 2.7) in expression.

Thus, we can proceed to a calculus on the space $(L_p^2)^-$. For more details and further developments see the literature[78].

Chapter 8

Innovation theory

The innovation approach is one of the most powerful methods of the study of stochastic processes and random fields when we discuss them in line with the white noise analysis starting from **Reduction**. Indeed, we may say that the main advantage of white noise theory is based on the use of the innovation, which naturally comes from the step of *reduction* (see Chapter 1, Section 1.2). Having had the step of reduction to have the innovation, we have been led to introduce classes of generalized functionals, the variables of which are consisting of idealized elemental variables, coming from the innovation, like white noise $\{\dot{B}(t)\}$ and Poisson noise $\dot{P}(t)$.

Because of the significance of the innovation, the present authors have published a monograph[71] to describe various results on random fields. It is therefore not wise to use too much pages on this topic; we shall, therefore, present only main topics briefly having a short history that may explain the idea of the study. Needless to say, some new results will be mentioned too.

Since a system of *idealized elemental random variables* has been introduced in our analysis, the innovation theory has made much progress. A short history presented below will be helpful to explain the original idea of the innovation approach.

Following our plan to develop white noise analysis from the step of reduction, we shall see that the innovation introduced in this chapter is most fitting for the analysis of random complex systems. We have unfortunately often met interesting processes and fields for which innovations do not exist or are difficult to be formed explicitly. We shall, as a second choice, introduce a notion of innovation in the *weak sense*. Such a topic will be developed here to some extent. Interesting applications of the results can be seen in the linear prediction theory. It is noted that for Gaussian processes innovation in the weak sense coincides with (real) innovation.

8.1 A short history of innovation theory

We shall not go into detail to the history of innovation theory. One may, however, just think of a story in the famous monograph "Ars Conjectandi" by J. Bernoulli that appeared in 1713 (after his death), where the word "stochastice" (stochastic) appeared for the first time. This notion lets us be awakened to innovation, but a quick review of what have been developed up to present does not look back upon such an old story. The reason for reviewing such an old history is that we hope to have the original idea of the present research rediscovered and that one can ask oneself if the present development does realize the original direction.

Now we will list only significant contributions to the innovation theory in a chronological order.

1. P. Lévy[102] (2nd ed. with supplement, 1964, Chapter 2) is essentially devoted to the discussion on how to define a stochastic process with special emphasis on the role of innovation, although he does not use the term "innovation" explicitly. We can however understand well what he intended.

   Note The innovation for the discrete parameter case was discussed in Lévy's book[101] Chapter VI, much earlier. The results should have somewhat different technique compared with the continuous parameter case. For one thing, the discrete parameter case does not need to require any continuity in time, so that the problem contains more freedom and hence becomes difficult.

2. Then follows P. Lévy's work[104].
   To define (or determine) a stochastic process $X(t)$, he provided a formal (intuitive) equation. It is expressed in the form for the variation $\delta X(t)$ of $X(t)$ over the infinitesimal time interval $[t, t+dt), dt > 0$F

$$\delta X(t) = \Phi(X(s), s \le t;\ Y(t), t, dt). \qquad (8.1.1)$$

It is called a *stochastic infinitesimal equation*, and it might be compared to an ordinary differential or variational equations. We should, however, recognize significant difference; actually equation (8.1.1) involves additional element $Y(t)$.

The $Y(t)$, which is usually an infinitesimal random variable, contains the new information which $X(t)$ gains during the infinitesimal time interval $[t, t+dt)$ and is independent of $X(s)$, $s \leq t$. In this sense, we may now call $Y(t)$ the *innovation*.

3. This idea of innovation was realized for Gaussian processes in the paper[35], in terms of the canonical representation. Section I of this paper is in agreement with the result of the paper by H. Cramér presented at the Fourth Berkeley Symposium on Mathematical Statistics and Probability held in 1960. He obtained the result independently. In reality, this concept of the canonical representation was proposed by P. Lévy in 1955 (his paper was published in 1956: Proceedings of the Third Berkeley Symposium). To obtain the innovation of a Gaussian process is the main part of finding canonical representation. This has been established as is shown in Section 3.2 although we did not use the term "innovation".

4. Before 1970, T. Kailath and his collaborators appealed to the innovation theory in mathematical theory of communication and engineering. Let $X(t)$ be the observed data which is assumed to be of the form

$$X(t) = s(t) + n(t), \tag{8.1.2}$$

where $s(t)$ is the signal and $n(t)$ is a noise. $s(t)$ is independent of the future noise $n(s)$, $s \geq t$.

Set $\hat{s}(t) = E(s(t)|\mathbf{B}_t(X))$. Then,

$$\nu(t) = X(t) - \hat{s}(t) \tag{8.1.3}$$

is the innovation of $X(t)$ given by (8.1.2). (See e.g. Kailath[82].)

It is interesting to note that A. N. Shiryaev obtained similar results around the same time, independently.

5. N. Wiener discussed linear and nonlinear (sample function-wise) prediction theory for stationary stochastic process. We now understand well the results by using the word innovation or canonical representation if the process is Gaussian. Later Wiener published the book[166], where Lectures 12 and 13 on Coding and Decoding are closely related

to the innovation theory.

6. G. Kallianpur used the term *innovation* in the note[83] titled "Some ramifications of Wiener's ideas on nonlinear prediction", and we are led to recall the joint work with Wiener "Nonlinear prediction" Technical Report 1, 1956, ONR (unpublished manuscript). In the note Kallianpur gives detailed interpretation on Wiener's ideas, in particular those facts described in Lectures 12 and 13, Coding, Decoding in Wiener's book mentioned above in Item 5, where the parameter is taken to be discrete, namely time series is discussed. We can see the ideas are close to those proposed by Lévy in his book[101], so far as innovation is concerned. The Wiener's work reviewed above continues to nonlinear prediction, which is one of the main subjects of his research.
7. As for the innovation of random fields, there are significant results by R. L. Dobrushin *et al*, and we have some general results on random fields in connection with the construction of the innovation. (See e.g. Si Si[143,144] and Hida-Si Si[71].)
8. The recent work by Accardi–Hida–Si Si[5] uses sample function techniques and also referred to the martingale theory developed by H. Kunita and S. Watanabe[92] to obtain the innovation.
9. Win Win Htay has reported the innovation approach to linear processes at the IIAS workshop organized by M. Ohya, 2003 (see the papers [153,167]).

Concerning a historical note on the innovation in the weak sense, we shall see later in Section 8.3.

8.2 Definitions and examples

We shall first introduce the notion of the innovation of a stochastic process and that of a random field in rather intuitive level.

One may note a definition of the innovation by a formal equation which is called a *stochastic infinitesimal equation* mentioned in Item 2 of the last section.

The equation suggests to us how to define the innovation, and eventually we will be led to construct a stochastic process in terms of the innovation.

The $\delta X(t)$ in formula (8.1.1) denotes the variation of $X(t)$ over an infinitesimal time interval $(t, t + dt), dt > 0$, and it is explained by the formula

$$X(t + dt) - X(t) = \delta X(t) + o(dt),$$

where the main part $\delta X(t)$ is usually of order $\sqrt{dt}$ in a suitable topology if new randomness appears at t.

The $Y(t)$, the innovation of $X(t)$, is a generalized stochastic process with independent values at every moment t in the sense of Gel'fand-Vilenkin[30], and it is independent of the $X(s), s \leq t$. The $Y(t)$ contains, as briefly glanced before, the same information as that is gained by $X(s)$ during the time interval $[t, t + dt)$ in terms of sigma fields of events.

With the spirit illustrated above we can finally come to a rigorous definition of an innovation of a stochastic process $X(t)$, for which separability is assumed.

Use the notation $\mathbf{B}_t(X)$, which denotes the sigma-field generated by $X(s), s \leq t$, in order to define the innovation $Y(t)$ in the following manner (see the literature[71] for more details):

Definition 8.1 $Y(t)$ is the innovation of $X(t)$ if the following conditions are satisfied.

1. $Y(t)$ is a generalized stochastic process with independent values at every moment,
2. $Y(\xi) = \langle Y, \xi \rangle, \xi \in E$ is measurable with respect to $\mathbf{B}_{t+}(X)$, where t is the supremum of the supp(ξ), the support of ξ, and independent of $\mathbf{B}_{s-}(X)$, where s is the infimum of the supp(ξ),
3. It holds that

$$\mathbf{B}_{t-}(X) \bigvee \left(\bigwedge_{t \in supp(\eta)} \mathbf{B}(Y(\eta)) \right) = \bigwedge_{\epsilon > 0} \mathbf{B}_{t+\epsilon}(X). \tag{8.2.1}$$

Once the innovation is obtained, the given process would be expressed in a formal equation

$$X(t) = \Psi(Y(s), s \leq t, t), \tag{8.2.2}$$

if there is no remote past, i.e. $\bigwedge_t \mathbf{B}_t(X)$ being the trivial field.

Note that t and s may be replaced by rational number, so that the measurability of each σ-field is guaranteed.

Thus, the pair $(\Psi, \{Y(t)\})$ completely determines the probabilistic structure of the $X(t)$, and hence, the given random complex system can be identified. Then, naturally follows the analysis, where $Y(t)$'s are taken to be the variables of the functional Ψ.

Needless to say, if $X(t)$ is complex valued, so is the innovation $Y(t)$.

We did not specify the values of $Y(t)$ explicitly; complex-valued, vector valued, even generalized stochastic process or infinite dimensional valued, etc. Every case is acceptable; we can adjust with necessary modification.

Example 8.1 Let a Gaussian process has generalized canonical representation, let it be denoted by (3.2.7). Then, $dY(t)$, according to the notation there, or $\dot{Y}(t)$ is the innovation. This happens in the case of higher multiplicity. It is, of course, not unique, but equivalent in the sense that is discussed in Theorem 3.3.

Example 8.2 (Hitsuda[73]) There is shown an example of a Gaussian process that has infinite multiplicity in a natural manner. The innovation is therefore infinite dimensional.

For a random field $X(C)$ parametrized by a $(d-1)$-dimensional manifold C in R^d, we shall extend the above definition. It is reasonable to assume that C is a smooth ovaloid, since we shall carry on variational calculus by taking C to be a variable. To express the idea explicitly, it is suitable, first of all, to propose a formal equation for a random field $X(C)$. If the C deforms slightly, say as much as δC, then $X(C)$ would vary as much as the variation $\delta X(C)$ ignoring the term $o(\delta C)$ with respect to a certain norm of C. The variation would be expressed in a formal expression:

$$\delta X(C) = \Phi(X(C'); (C') \subset (C), Y(s), s \in C, C, \delta C), \tag{8.2.3}$$

in which (C) denotes the domain enclosed by C and where $\{Y(s), s \in C\}$ is the innovation. This is a generalization of the infinitesimal equation proposed by P. Lévy. With such a formal equation in mind, we are going to give a definition of the innovation for $X(C)$. To make the statement rigorous, we prepare some notations.

Let $\mathbf{B}_C(X)$ be the sigma-field generated by all the $X(C')$ with $C' < C$, that is, the domain inside of C' is a subset of the domain inside of C : $(C') \subset (C)$. Similarly, $\mathbf{B}_C(Y)$ is defined in such a way that

$$\mathbf{B}_C(Y) = \bigwedge_{supp(\xi)\supset(C)} \mathbf{B}(Y(\xi)). \tag{8.2.4}$$

Definition 8.2 The innovation of $X(C)$ is a family of systems $Y(C) = \{Y(s), s \in C\}$, with $C \in \mathbf{C}$ such that

i) $Y(s)$ is a generalized stochastic process, may be a vector-valued process, parametrized by $s \in C$ and has independent values at every s,
ii) $Y(\xi), \xi \in E$, is independent of $\mathbf{B}_C(X)$ if $supp(\xi)$ is disjoint with (C'), and if $Y(\xi)$ is $\mathbf{B}_C(X)$-measurable for ξ with $supp(\xi) \subset (C)$.
iii) It holds that

$$\mathbf{B}_{C-}(X) \bigvee \left(\bigwedge_{\xi} \mathbf{B}(Y(\xi)) \right) = \bigwedge_{\delta n > 0} \mathbf{B}_{C+\delta C}(X), \tag{8.2.5}$$

where $\mathbf{B}_{C-}(X)$ is the sigma-field generated by $\mathbf{B}_{C'}(X)$ such that the closure of C' is a subset of (C), and where ξ satisfies $supp(\xi) \supset (C)$.

Let a manifold C be parametrized by s, i.e. $C = \{s\} \subset R^d$.
We understand that the deformation δC of C is represented by a system

$$\delta C \simeq \{\delta n(s), s \in C\}, \tag{8.2.6}$$

where $\delta n(s)$ denotes the length of the normal vector to C at s.

For each case $X(t)$ or $X(C)$, the numerical values of the innovation are not unique, however we can prove the following:

Triviality. In both cases, stochastic processes and random fields, the sigma-field generated by the innovation is unique, assuming that the innovation exists.

Applications of innovation theory for random fields will be discussed in the next chapter.

8.3 Innovations in the weak sense

The theory of the innovation in the weak sense, that we are going to explain in this section, often plays an important role in stochastic analysis. In particular, the theory is useful in the case where linear prediction is the main purpose. More particularly, the (nonlinear) predictor of Gaussian processes coincides with the linear predictor, so that the innovation in the weak sense provides enough information to get the predictor. Or, we often meet the case where the innovation is not easy to obtain, but we may substitute the innovation in the weak sense by the proper innovation.

Let $X(t,\omega), t \in T,\ \omega \in \Omega(P)$, be a *second order process*, namely $E(|X(t)|^2 < \infty)$ and $X(t)$ is "mean continuous":

$$E|X(t+h) - X(t)|^2 \ \rightarrow\ 0, \text{ as } h \rightarrow 0.$$

To fix the idea, we assume that $T = R$ and $E(X(t)) = 0$.

By assumption, $X(t)$ is viewed as a continuous curve in the Hilbert space $L^2 = L^2(\Omega, P)$. Let $M_t(X)$ be the linear subspace of (L^2) spanned by $X(s), s \leq t$. Then, $\{M_t(X), t \in R\}$ is an increasing family of closed subspaces of L^2. Since $X(t)$ is mean continuous,

$$M_{t+}(X) = \bigwedge_{\epsilon>0} M_{t+\epsilon}(X),$$

is equal to $M_t(X)$.

Set

$$M(X) = \bigwedge_t M_t(X),$$

which is the subspace of L^2 spanned by all $X(s), s \in R$.

Now we must assume a basic property of the $X(t)$ such that it is "purely non-deterministic". Sometimes this requirement means that $X(t)$ has "no remote past". This property is expressed in the equation

$$\bigwedge_t M_t(X) = \{0\}. \tag{8.3.1}$$

Associated with the subspace $M_t(X)$ is a projection operator $E(t)$ such that

$$E(t) : M(X) \mapsto M_t(X).$$

The following assertion is obvious.

Proposition 8.1 *The assumption that a second order process $X(t)$ is purely non-deterministic implies that the family of $\{E(t), t \in R\}$ of projections on $M(X)$ is a resolution of the identity operator I.*

Since we shall be concerned with linear functionals of $X(t)$'s, we are in a similar situation to Section 3.2, where Gaussian processes are dealt with. So far as linear computations are discussed by using Hilbert spaces spanned by the $X(t)$'s, we can proceed to the calculus as in the Gaussian case. We can therefore appeal to the Hellinger-Hahn theorem again to obtain a direct sum decomposition of $M(X)$ into cyclic subspaces expressed in the following form:

$$M(X) = \bigoplus_{n=1}^{N} M_n(X_n), \ 1 \le N \le \infty,$$

for a nontrivial process $X(t)$. In this direct sum, the spectral measures $E(|dE(t)X_n|^2) = d\rho_n(t)$ are arranged in a decreasing order, and N is the *multiplicity* of $X(t)$. The notion of multiplicity is given for the second order processes, not necessarily to be restricted to Gaussian processes.

Formally speaking, $dE(t)X_n = dZ_n(t)$ with $dt \ge 0$, is orthogonal to $M_t(X)$ and it defines an *orthogonal* random measure.

Definition 8.3 The collection $\{dZ_n(t)\}$ is called a *system* of innovations of $X(t)$ in the weak sense.

Definition 8.4 If, in particular, $N = 1$, then the $dZ_1(t)$, denoted simply by $dZ(t)$, is called an *innovation* of $X(t)$ *in the weak sense.*

Triviality 1 If an innovation in the weak sense exists, it is uniquely determined up to multiplication by non-zero, non-random function.

Triviality 2 For a Gaussian process, the innovation in the weak sense coincides with the innovation defined originally.

We now focus our attention on a second order *stationary* stochastic process $X(t), t \in R$, with $E(X(t)) = 0$.

The mapping

$$U_t : X(s) \mapsto X(s+t), \;\; s \in R,$$

extends to a unitary operator on $M(x)$. The relationships

$$U_t U_s = U_s U_t = U_{t+s}$$

are obvious, and U_t is strongly continuous in t :

$$U_{t+h} \to U_t, \text{ strongly in } M(X) \text{ as } h \to 0.$$

Hence Stone's theorem asserts the following.

Theorem 8.1 *The U_t has the spectral representation*

$$U_t = \int e^{it\alpha} dG(\alpha) \tag{8.3.2}$$

where $G(\alpha)$ is a projection operator and $\{G(\alpha), \alpha \in R\}$ is a resolution of the identity.

Corollary 8.1 *The $X(t)$ has the spectral decomposition of the form*

$$X(t) = \int e^{it\alpha} dZ(\alpha) \tag{8.3.3}$$

where $dZ(\alpha)$ is an orthonormal random measure.

Proof. By Theorem 8.1, we have

$$X(t) = U_t X(0) = \int e^{it\alpha} dE(\alpha) X(0).$$

Set $dZ(\alpha) = dE(\alpha)X(0)$. Then it is easy to show that $\{dZ(\alpha)\}$ is an orthonormal random measure.

■

As a result, we obtain the spectral representation of the covariance function $\gamma(h)$:

$$\gamma(h) = E(X(t+h)X(t)) = \int e^{ih\alpha} dm(\alpha), \tag{8.3.4}$$

where $dm(\alpha) = E|dE(\alpha)X(0)|^2$. which is called the spectral measure.

Assume further that $X(t)$ is purely non-deterministic. Then it is known (e.g. see the literature[84]) that $dm(\alpha)$ is absolutely continuous and its density $f(\alpha) = \frac{dm(\alpha)}{d\alpha}$ satisfies

$$\int \frac{|\log f(\alpha)|}{1+\alpha^2} d\alpha < \infty.$$

Then, there exists a unique (up to equivalence) factorization $F(\alpha)$ exists and it vanishes on $(-\infty, 0]$. Finally, we have

Theorem 8.2 *Under the above assumptions, the $X(t)$ has unit multiplicity and is expressed in the form*

$$X(t) = \int_{-\infty}^{t} F(t-u)dY(u), \tag{8.3.5}$$

where $F(u)$ is a canonical kernel and $dY(u)$ is a normalized (i.e. $E(|dY(u)|^2) = du$) orthogonal random measure which is the innovation in the weak sense. The covariance function $\gamma(h)$ of $X(t)$ is given by

$$\gamma(h) = \int_{-\infty}^{0} F(h-u)F(-u)du, \ h > 0.$$

Remark 8.1 *A canonical kernel $F(t,u)$ mentioned above is understood to be the same as in the case of Gaussian process. We now understand that the representation by the integral with the kernel $F(t,u)$ implies the equality*

$$M_t(X) = M_t(Z) \quad \text{for every } t.$$

One may ask that if the following assertion is true. If, in the above equality, the $M_t(X)$ is replaced by $L_t^2(X) = L^2(\Omega, \mathbf{B}_t(X), P)$, with $\mathbf{B}_t(X)$ the σ-field determined by the $X(s), s \le t$, and if $M_t(Z)$ is replaced by the space of nonlinear functions of Z, then the innovation in the weak sense, which always exists, could be equal to the (original) innovation. The answer is not true. This fact can be illustrated by the following example.

Example 8.3 Let

$$X(t) = B(t)^2 - t.$$

Then, the infinitesimal random variable $B(t)dB(t), dt > 0$, is orthogonal to $L_t^2(X)$. The $X(t)$ has a cyclic vector $Z_1(t)$ such that $B(t)dB(t) = dZ_1(t)$. But it is not independent of $L_t^2(X)$.

Another example of an innovation in the weak sense illustrates a connection with the Fourier series.

Example 8.4 (discrete parameter analogue).

On the probability space $(S^1, \frac{d\theta}{2\pi})$, we define

$$X_n = e^{in\theta}, \quad n \neq 0,$$
$$X_0 = 0.$$

The innovation in the weak sense is X_n itself. However, if the case where nonlinear functions are taken, then we have almost trivial result; indeed, X_n is deterministic.

An interesting generalization of Theorem 8.2 can be seen in K. Itô's paper that discusses stationary random distributions[77] which are weakly stationary generalized stochastic processes.

8.4 Some other concrete examples

Good examples for which innovations can actually be obtained are now in order. We have, however, to note that there are interesting processes for which innovation does not exist or cannot be obtained explicitly; this is really unexpected.

Typical cases are listed below, although some of them need more or less to be explained.

1) Gaussian processes and Gaussian random fields.

See Chapter 3, Section 3.2, where the canonical representation theory of Gaussian processes is discussed. Unit multiplicity for the process means existence of the simple innovation. In general, if a process is separable and purely non-deterministic, then we are given an innovation which is multi-dimensional.

As for Gaussian random fields, we can discuss as in Section 8.2. If the random field is simple Markov, then we can form the innovation explicitly. Detailed discussions are given in the monograph[71], Chapter 5.

2) Second order stationary processes.

Second order stationary processes that are mean-continuous and have no remote past always have their innovations in the weak sense, as was discussed in Section 8.3.

3) Linear processes and linear fields.

Details are discussed in Chapter 4. Among others, innovation problem has been dealt with through the representation theory for linear processes.

In general, a linear process has components, one is Gaussian and the other is a linear functional of a compound Poisson process. The innovation is therefore a combination of those two parts. Significant difference between the components Gaussian and non-Gaussian is that innovation is to be constructed by observing sample functions in the latter case. As for the Gaussian component we can use Hilbert space technique, while the Poisson case is not like that.

4) Behavior of sample functions.

N. Wiener's results with emphasis on the behavior of sample functions in the study of prediction and nonlinear functions of a process. His idea has close connection with the innovation approach. See the monographs[165,166], where innovation can be seen in the course of obtaining the predictors. Also, see the paper[83] to see that significance of the innovation is extensively explained.

5) Processes that are linear in homogeneous chaos.

If a process is expressed in terms of single homogeneous chaos or multiple Wiener integrals of a certain degree, then we may obtain analogous results to the Gaussian case, i.e. linear operations are used to get the innovation, although it is taken to be in the weak sense. However, the results are useful in the prediction theory and in other applications.

6) Some martingale; processes and fields

Kunita-Watanabe theory (see the literatures[92]) to see the details, where we can see that the martingale property plays the key role to obtain the innovation.

7) Multiple Markov Gaussian processes and fields.

For the canonical representations of multiple Markov Gaussian processes, we know that the kernels are Goursat kernel of finite order. We can therefore propose concrete construction of the innovation by using analytic methods. For this purpose, the results in Section 3.3 may be slightly modified. The modification is almost straightforward, so is omitted here. Concerning the fractional Brownian motion we can also obtain the innovation explicitly as was discussed in Section 3.4.

8) Stationary processes with the help of the time operator.

We are interested in the relationship between innovation and time operator. We shall be back to this problem in Chapter 10. Having the results there, one can proceed to the actual method of getting innovation.

And so forth.

Remark 8.2 *Innovation may be formed by a single elemental additive process (non-decomposable innovation). There are, however, cases where many additive processes are involved. This situation may require decomposability of the innovation and it can be dealt with in connection with the Lévy decomposition of additive processes.*

Example 8.5 This example does not claim to be included in the above list, but it is worth to be mentioned here so that it will become a general important problem.

Let $X(t), t \geq 0$, be given by

$$X(t) = P_{u_1}(t) + P_{u_2}(t),$$

where $P_{u_i}(t), i = 1, 2$, are mutually independent Poisson processes with scales of jump $u_i, i = 1, 2$, respectively. If the ratio u_1/u_2 is irrational, then, observing sample functions locally in time, $X(t)$ can be decomposed into two Poisson processes as is easily seen, and the two are taken to be a two-dimensional innovation. Note that preliminary knowledge on $X(t)$ is that it is a sum of two independent Poisson processes with the ratio of scales of the jumps is being irrational.

If the ratio is rational, then making global observation of sample functions in the past, we can still decompose sample functions into two paths.

Remark 8.3 *The whitening of a completely linearly correlated system discussed in Chapter 3 may be considered as a generalization of the innovation problem.*

Chapter 9

Variational calculus for random fields and operator fields

9.1 Introduction

Random field $X(C)$ parametrized by a manifold C and its variational calculus have been discussed in the monograph[71] published in 2004. There we emphasized the importance of the innovation approach to the study of stochastic processes and random fields. We are always in this line, and by using this method we wish to shed light on complex dependence of the $X(C)$ when C runs through a certain class $\mathbf{C}$, prescribed below, of manifolds in a Euclidean space.

For this purpose we shall review the known theorems on variational calculus quickly and discuss basic ideas by illustrative examples. The class of random fields in question is somewhat wider than what we have dealt with in the monograph[71].

Let $X(C)$ be a random field in $(L^2)^-$ on S^*, where the parameter C, which is a $(d-1)$-dimensional manifold, runs through a class $\mathbf{C}$, defined by

$$\mathbf{C} = \{\text{ovaloid } C : C \in \mathbf{C}^\infty\},$$

where we tacitly assume that the usual Euclidean norm is used. The infinitesimal deformation of C is denoted by δC as in (8.2.6). For such an infinitesimal deformation we are given a variation $\delta X(C)$ of $X(C)$. We can therefore define a so-to-speak *stochastic variational equations*. The equations in question are similar to stochastic differential equations, however they should be dealt with in more complicated manner. To avoid complex treatments, what we shall discuss in this chapter are much restricted.

9.2 Stochastic variational equations

We are going to discuss stochastic variational equations under the restrictions of the form explained below. Having been suggested by the illustrated examples that will appear in Section 9.3, namely Examples 9.1–9.4, we shall consider the case where the equations to be discussed are expressed in the form (Volterra form) :

$$\delta X(C) = \int_{s \in C} X'(C,s)\delta n(s)ds, \tag{9.2.1}$$

where $X'(C,s)$ is a random field depending on $(C,s), C \in \mathbf{C}, s \in C$.

Suppose $X'(C,s)$ is in $(S)^*$ for any pair (C,s).

Now let us remind the S-transform of white noise functional (may be generalized functional) $\varphi(x), x \in E^* \subset L^2(R^d)$. We are concerned with multi-dimensional parameter white noise, and the S-transform has the same expression as the case $d = 1$.

$$S : \varphi \mapsto (S\varphi)(\xi) = \exp[-\frac{1}{2}\|\xi\|^2] \int_{E^*} \exp[\langle x, \xi \rangle]\varphi(x)d\mu(x).$$

Applying the S-transform to $\delta X(C)$, we have

$$\delta U(C) = \int_I U'(C,s)\delta n(s)ds, \tag{9.2.2}$$

where C is parametrized by $s \in I = [0,1]$, and where $U(C) = S(X(C))$.

Letting C be represented by a function $\xi(s), s \in S^1$, we are given, by identifying $U(C) = U(\xi)$, the following form of variation

$$\delta U(\xi) = \int_I f(\xi, U, s)\delta n(s)ds, \tag{9.2.3}$$

where ξ is assumed to be a member of a nuclear space E. This equation is the so-called *Volterra form* for variational equation. We shall hereafter deal with functionals such that the variations of the functionals are expressed in the Volterra form.

Note. Equation (9.2.3) may be considered as a generalization of total differential equation

$$du = \sum_{i=1}^{n} p_i(x)dx_i, \quad x = (x_1, x_2, \cdots, x_n) \in R^n.$$

In this sense, P. Lévy[103] proposed variation of the form (9.2.3). Under some assumptions, he proved the existence and the uniqueness of the solution.

Once the solution $U(\xi)$ is obtained, it is necessary to check if $U(\xi)$ is an image of some white noise functional under S-transform. For this purpose the Potthoff-Streit[64,126] criterion is useful. In this case we should note that the theorem is true for any functional representation of C.

9.3 Illustrative examples

Example 9.1 Additive Gaussian random field.

Let $x(u), u \in R^2$, be a two-dimensional parameter white noise and let $\mathbf{C}$ be a class of smooth ovaloid in R^2. Define a Gaussian random field $X(C), C \in \mathbf{C}$, by

$$X(C) = \int_{(C)} f(u)x(u)du, \quad (C_0) \subset (C), \tag{9.3.1}$$

where $f(u)$ is a non-random function belonging to the Sobolev space over R^2 of order 1 and never vanishes. We put initial data $X(C_0)$.

The $X(C)$ is additive in (C) in the sense that if $(C_1) \subset (C_2)$, then $X(C_2) - X(C_1)$ is independent of any $X(C')$ with $(C') \subset (C_1)$. Such a random field can be characterized by the stochastic variational equation of the form

$$\delta X(C) = \int_C f(s)x(s)\delta n(s)ds, \tag{9.3.2}$$

where $C + \delta C$ is chosen outside of C.

Note that by the assumption on $f(u)$ the value $f(s)x(s)$ is well defined and integrable over C. Then, we apply diffeomorphism of C which determine the values $f(s)x(s)$ as a generalized white noise functional on C.

The S-transform of (9.3.2) is of the form (9.2.3), so that the solution exists. In fact, we have (9.3.1), and $\delta n(s)$ is arbitrary, so that $x(s)$ can be obtained since $f(s)$ can be obtained as $f(s) \neq 0$.

Thus, we are given the innovation of $X(C)$. Although it is not normalized, it is not important because $f(s)$ does not vanish.

Example 9.2 The Langevin type variational equation.

Assume that $X(C)$ has a variation $\delta X(C)$ expressed in the form :

$$\delta X(C) = \delta f(C) \int_{(C)} g(u)x(u)du + f(C) \int_C g(s)x(s)\delta n(s)ds, \qquad (9.3.3)$$

where $f(C)$ is smooth and never vanishes and where $g(s)$ is continuous. The initial data is $X(C_0) = 0, (C_0) = \{0\}$.

To solve the equation we set

$$Y(C) = \frac{X(C)}{f(C)}.$$

Then it is easy to see that $\delta Y(C)$ satisfies the same equation as (9.3.2), so that we can play the same game as in Example 9.1.

Example 9.3 Functionals given by an integral over C.

Lévy's functional analysis[103] (Part I, §13) gives us a variational formula for a functional $I = \int_C uds$:

$$\delta I = \int_C (\frac{du}{dn} - \kappa u)\delta n ds, \qquad (9.3.4)$$

where $\frac{du}{dn}$ denotes the normal derivative and where κ is the curvature.

Example 9.4 A random field $X(C)$ given by

$$X(C) = \exp[\int_{C)} g(u)X(u)du]$$

with continuous function g satisfies (9.2.3). Detailed discussion is omitted.

We can apply this formula to the U-functional associated with the white noise functional expressing the conditional expectation arising from R^2-parameter Lévy's Brownian motion $X(a), a \in R^2$ (See the literature[139]). We recall Definition 2.1 that defines Lévy's Brownian motion.

Let $S(t), t > 0$, be a circle with center O (origin) and radius t. Suppose the values of a Lévy's Brownian motion $X(a)$ are known for all $a \in S(t)$.

Then, we wish to know the conditional expectation of $X(p), p$ being inside the $S(t)$. The exact expression of the conditional expectation $X(p|t) = E(X(p)|X(t\theta), 0 \leq \theta < 2\pi)$ is given as follows:

$$X(p|t) = \int_{S(t)} f(p, \theta, t) X(t\theta) d\theta, \tag{9.3.5}$$

where

$$f(p, \theta, t) = \frac{(t^2 - x^2)^2}{8t\rho(x, t, \theta - \beta)} + \frac{1}{2\pi}\left(1 - \frac{t+x}{2t}\mathbf{E}(\frac{\pi}{2}. \frac{2\sqrt{tx}}{t+x})\right), \tag{9.3.6}$$

the symbol $\mathbf{E}$ means the elliptic function.

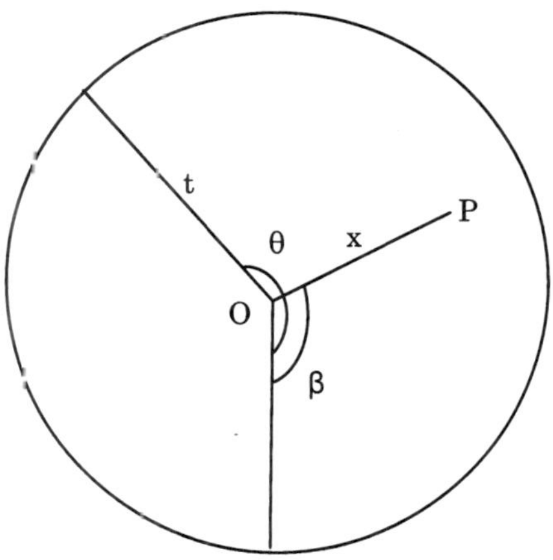

M(θ=0)

Fig. 9.1.

In the above formula we use the notations as are shown in Fig. 9.1. Namely, x is the distance between p and a fixed point m on $S(t)$, θ is the angle measured from m to the point $t\theta$ on $S(t)$, β is the $\angle pOm$, $\mathbf{E}$ is the elliptic function, and finally

$$\rho(x, t, \theta) = \sqrt{(x^2 + t^2 - 2xt\cos\theta)}.$$

We are interested in the variation of $X(p/t)$ by letting $S(t)$ vary in the class $\mathbf{C}$ the set of concentric circles; namely $\mathbf{C} = \{S(t), t > 0\}$. We have

$$\delta X(p|t) = \frac{d}{dt} X(p|t), \tag{9.3.7}$$

where the Lévy's formula (9.3.4) is used by taking $\frac{d}{dn} = \frac{d}{dt}$ and $\kappa = \frac{1}{t}$. More precisely,

$$\delta X(p|t) = \int_{S(t)} \left\{ \frac{d}{dt}(f(p,\theta,t)X(t\theta)) - \frac{1}{t} f(p,\theta,t)X(t\theta) \right\} \delta n d\theta.$$

Remark 9.1 *In the original paper[139] the point m, in the above diagram, is taken to be the origin, however for the sake of taking variation, origin is moved to o. Such a translation of parameter never changes the expression of the formula of the conditional expectation.*

9.4 Integrals of operators

In Chapter 5 we have discussed Laplacians and number operators that are expressed as quadratic forms of operators ∂_t (annihilation) and ∂_t^* (creation) where t runs through an interval of R^1. We now consider some more general operators that are linear and quadratic forms of those creation and annihilation operators.

9.4.1 *Operators of linear form*

1. Operator $\partial_t + \partial_t^*$

Multiplication by the variable $x(t)$ is expressed as

$$x(t) \cdot = \partial_t + \partial_t^* \tag{9.4.1}$$

as is well known.

Remark 9.2 *It is interesting to note that there are methods to let the multiplication $\partial_t + \partial_t^*$ be separated to get a pair $\{\partial_t, \partial_t^*\}$. This fact suggests white noise quantization.*

2. Operator fields

Let $\mathbf{C}$ be a class of smooth ovaloids in R^2. Take $C \in \mathbf{C}$ and consider

$$A(C) = \int_C f(\theta)\partial_\theta d\theta$$

and

$$A^*(C) = \int_C f(\theta)\partial_\theta^* d\theta,$$

where C is a smooth ovaloid, as usual, that is parametrized by $(u(\theta), v(\theta))$, $\theta \in [0, 2\pi]$, and where f and g are smooth functions. The domains of $A(C)$ and $A(C)^*$ are rich enough in (S) and $(S)^*$, respectively.

Example 9.5 Let $\varphi(x)$ be given by

$$\varphi(x) = \int\int F(u, v) : x(u)x(v) : dudv.$$

Then we have, noting $F(u, v) = F(v, u)$,

$$A(C)\varphi(x) = 2\int_C d\theta \int F(u(\theta), v)x(v)dv.$$

Example 9.6 Let $A^*(C)$ acts on the constant 1. Then we are given a Gaussian field depending on $C \in \mathbf{C}$ expressed on the form

$$A^*(C)1 = \int_C x(\theta)d\theta.$$

9.4.2 *Operators of quadratic forms of the creation and the annihilation operators*

1. Rotations

As was briefly mentioned in Section 5.8, the infinitesimal generator $r_{s,t}$ of operators acting on white noise functionals coming from the rotations on $(x(s), x(t))$-plane is given by

$$r_{s,t} = \partial_s^* \partial_t - \partial_t^* \partial_s.$$

One can see that this is an analogous form of the infinitesimal rotation acting on $L^2(R^d)$, since $r_{s,t}$ may be rephrased as

$$r_{s,t} = x(s) \cdot \partial_t - x(t) \cdot \partial_s,$$

which implies the above formula by using the expression (9.4.1).

As is expected, it holds that

$$r_{s,t} = -r_{t,s}.$$

Note that both Volterra (Gross) and Lévy Laplacians commute with the $r_{s,t}, s, t \in R$.

It is now the time to give some reasonable interpretation to the expression of the Volterra Laplacian Δ_V and Lévy Laplacian Δ_L discussed in Section 5.8, where the parameter set is restricted to $[0, 1]$, for convenience. The parameter set can easily be generalized to general interval, including the entire R. It satisfies the following properties which are to be expected :

i) it is quadratic in ∂_t's,
ii) it commutes with rotations $r_{s,t}$,
iii) it annihilates some subspace of (L^2) (or (S)), and
iv) it is negative.

As for the property iii), the subspace is taken to be H_0 involving only constant for Δ_V, and (L^2) for Δ_L. This fact shows an estrangement between these two Laplacians.

We continue to give interpretation on the two Laplacians. It is useful for the characterization and for applications.

We consider quadratic forms $Q = Q(\partial_u, \partial_v)$ of the annihilation operators expressed in the form

$$\int\int F(u, v)\partial_u \partial_v du dv,$$

where the kernel function $F(u, v)$ is symmetric and can be a generalized function, specified depending on the situations.

First, we discuss commutation relations with the rotations.

Proposition 9.1 *For any $r_{t,s}$ we have*

$$[Q(\partial_u, \partial_v), r_{t,s}] = 2\partial_s \int F(t, u)\partial_u du - 2\partial_t \int F(s, v)\partial_v dv. \tag{9.4.2}$$

Proof. We note that $[\partial_t^*, \partial_s] = -\delta(t - s)$. Then, by actual computation, we can prove

$$[\partial_u \partial_v, \gamma_{t,s}] = \delta(t-u)\partial_v \partial_s^* + \delta(t-v)\partial_u \partial_s - \delta(s-u)\partial_v \partial_t - \delta(s-v)\partial_u \partial t.$$

This implies the conclusion of the proposition.

■

Proposition 9.2 *If the quadratic form $Q(\partial_u, \partial_v)$ commutes with every $r_{t,s}$, then Q is the Volterra Laplacian up to constant.*

Proof. By assumption the formula (9.4.2) vanishes:

$$\partial_s \int F(t,u)\partial_u du - \partial_t \int F(s,v)\partial_v dv = 0.$$

Hence

$$F(t,u) = \delta_t(u).$$

This shows that $Q = \Delta_V$.

■

As for a characterization of the Lévy Laplacian we must take the subspace (L^2) to be annihilated and add one more condition that

v) it should give the trace of quadratic functional in $H_2^{(-2)}$ up to constant 2.

In this case we have to claim formally

$$F(t,u) = (\delta_t(u))^{-1} dt,$$

so that we are given the desired expression

$$\Delta_L = \int \partial_t^2 (dt)^2.$$

Computations behind are

$$\int\int \left(\frac{1}{\delta(t-s)}\partial_t \partial_s\right) \delta(t-s) ds dt = \int \frac{1}{\delta(0)}\partial_t^2 dt = \int \partial_t^2 (dt)^2.$$

9.4.3 *Polynomials in* $\partial_t, \partial_s^*; t, s \in R$, *of degree 2*

First we have to explain why we stick to polynomials in the creation and annihilation operators of degree 2. First we recognize fundamental roles of the rotations $\gamma_{s,t}$ which are quadratic and satisfy $\gamma_{s,t}^* = \gamma_{t,s}$.

Next we are interested in operators in similar forms and have good relationships with $\gamma_{s,t}$. The relationships may be expressed in terms of the Lie product, so that we consider a Lie algebra generated by operators in question. We assume the algebras are finitely generated on their direct sum.

There is one more important reason. A well-known algebra called Schrödinger algebra has good connection with what we have in mind. The Lie algebra is

$$A = \left\{\delta, \partial_t, \partial_t^*, (\partial_t)^2, (\partial_t^*)^2, n_t = \partial_t\partial_t^*, n_t^*, t \in R\right\}.$$

Since the delta function δ is involved, the algebra is understood in a generalized sense.

As is easily seen, A is closed under the Lie product [,], which means it is a Lie algebra. Obviously A commutes with rotations $r_{s,t}$.

On the other hand, an operator of degree 3, say ∂_t^3 is added to A, then the algebra generated by A cannot be finite system.

Now we consider

General quadratic forms.

Let D be a domain of R^2 such that its boundary is an ovaloid or D is a square. Set

$$B^{i,j}(D) = \int_D Q^{i,j}(u,v)dudv, \quad i,j = 0 \text{ or } 1,$$

where

$$Q^{0,0}(u,v) = a^{0,0}(u,v)\partial_u\partial_v \tag{9.4.3}$$

$$Q^{0,1}(u,v) = b^{0,1}(u,v)\partial_u\partial_v^* \tag{9.4.4}$$

$$Q^{1,0}(u,v) = c^{1,0}(u,v)\partial_u^*\partial_v \tag{9.4.5}$$

$$Q^{1,1}(u,v) = d^{1,1}(u,v)\partial_u^*\partial_v^*, \tag{9.4.6}$$

the kernel functions may be taken to be functions in the Sobolev $K^{-3/2}(R^2)$.

We are specifically interested in the commutation relations among $B^{i,j}$'s as well as those with the rotations. Note the characterization of the Lévy Laplacian by requiring the property that it commutes with rotations. Similar commutative property can be seen in $B^{1,1}(D)$ with D being a square.

We also note the collection of the operators of linear form and quadratic forms with D being a square forms a Lie algebra. Beyond quadratic forms, say a collection of polynomials in annihilation operators and creation operators of degree three and higher, we cannot expect a (closed) Lie algebra.

Chapter 10

Four notable roads to quantum dynamics

This chapter is devoted to evaluation of four notable roads that white noise analysis aims to achieve within the fields of quantum dynamics. The first is Feynman's path integral and the second is an infinite dimensional Dirichlet form. The third one is concerned with time operator, which is rather classical topic, however at present a new viewpoint can be given. In this sense, it is worth mentioning on this occasion. Finally, the fourth road is towards Euclidean fields and unitary representations of quantum field which will be touched upon only briefly.

10.1 White noise approach to path integrals

The idea of our method for path integrals in quantum dynamics is to utilize the white noise measure to establish an infinite dimensional integration on function spaces. This method, as is well known, has been originated by R. Feynman, with some motivation due to Dirac's idea. It is viewed as a third method of quantization, which is different from the formulation by W. Heisenberg and the one by E. Schrödinger. Our method of path integral within the frame work of white noise analysis follows mainly the Feynman's method in spirit. However, some other quantum mechanical considerations are taken into account.

In quantum dynamics there are many possible trajectories of a particle, and each trajectory may be viewed as a sum of the classical one and fluctuation. Our assertion is that the amount of the fluctuation should be expressed as a Brownian bridge. To explain the reason why a Brownian bridge is fitting for a realization of fluctuation, it seems better to illustrate some particular properties and characteristics of a Brownian bridge.

First, we need to give a characterization of a Brownian bridge which is denoted by $X(t), t \in [0,1]$, over the unit time interval, in order to see why it plays a dominant role in our setup of the path integral. It has to describe the amount of fluctuation around a classical trajectory; which is uniquely determined by the Lagrangian.

Let us review that a Brownian bridge is a Gaussian 1-ple Markov process with mean 0 and with covariance function

$$E(X(t)X(s)) = \Gamma(t,s) = (t \wedge s)(1 - t \vee s), \quad s,t \in [0,1],$$

(see Section 3.3).

Heuristically speaking, it was 1981 when we proposed a white noise approach to path integrals to have quantum mechanical propagators (see Hida-Streit[72], Streit-Hida[160]). We had, at that time, some idea in mind for the use of a Brownian bridge, and we had practically many good examples of integrand with various kinds of potentials, and satisfactory results have been obtained.

The original idea of employing a Brownian bridge came from some physical intuition (see the literature[71]) and we should now explain why a Brownian bridge is suitable from a mathematical viewpoint. Our idea can be generalized to the case where general parameter set appears, like multi-dimensional space or Lie algebra as we shall see later.

We now have a theorem

Theorem 10.1 *The Brownian bridge $X(t)$ over the time interval $[0,1]$ is characterized (up to constant multiplication) by the following conditions*

i) $X(t)$ is a Gaussian 1-ple Markov process with no initial data (so that it has the canonical representation),
ii) $X(0) = X(1) = 0$ (bridged), and $E(X(t)) \equiv 0$,
iii) the normalized process $Y(t)$ enjoys the projective invariance,

Proof. Assumptions i) - iii) proves that the covariance function $\Gamma(t,s)$ of the process to be determined has to be of the form

$$\Gamma(t, s) = \frac{f(\frac{s}{1-s})}{f(\frac{t}{1-t})}. \tag{10.1.1}$$

Further, the assumption i) asserts that $X(t)$ has the same continuity of a Brownian motion. Hence, the function f has to be proportional to the square root of the formula in (10.1.1). Thus, the theorem is proved.

■

The Brownian bridge just determined above has a canonical representation and is expressed in the form (up to constant):

$$X(t) = (1-t)\int_0^t \frac{1}{(1-u)}\dot{B}(u)du. \tag{10.1.2}$$

The covariance function $\Gamma(t, s)$ of the normalized process $Y(t)$ where

$$Y(t) = \frac{X(t)}{\sqrt{\Gamma(t,t)}} \tag{10.1.3}$$

is of the form

$$\Gamma(t, s) = \sqrt{(0, 1; s, t)}, \tag{10.1.4}$$

where $(0, 1; s, t)$ denotes the anharmonic ratio, that is $\frac{(1-t)/t}{(1-s)/s}$.

Note 1. We have required, as the assumption, the existence of the canonical representation as above. This requirement is essential, although the assumption may not look like a key condition.

Note 2. Characterization of a Brownian bridge, like Theorem 10.1, is useful when path integral type average is required for a dynamical system with a general parameter space. In such a case it is required to introduce a suitable fluctuation to the classical expression of the dynamics. For example, we shall see this fact on Chern-Simons action integral.

By the expression of the covariance function, it is easy to see that a Brownian bridge is reversible in time. As was mentioned before, our method of establishing a rigorous Feynman path integral is influenced by the idea due to Dirac and Feynman. This fact will be illustrated probabilistically in what follows.

The actual expression and computations of the propagator are given successively as follows:

We follow the Lagrangian dynamics. The possible trajectories are sample paths $y(s), s \in [0,1]$, expressed in the form

$$y(s) = x(s) + \sqrt{\frac{\hbar}{m}} B(s), \tag{10.1.5}$$

where the $B(t)$ is an ordinary Brownian motion. Hence the action S is expressed in the form in terms of quantum trajectory y:

$$S = \int_0^t L(y(s), \dot{y}(s)) ds.$$

Note that the effect of forming a bridge is given by putting the delta-function $\delta_0(y(t) - y_2)$ as a factor of the integrand, where $y_2 = x(t)$.

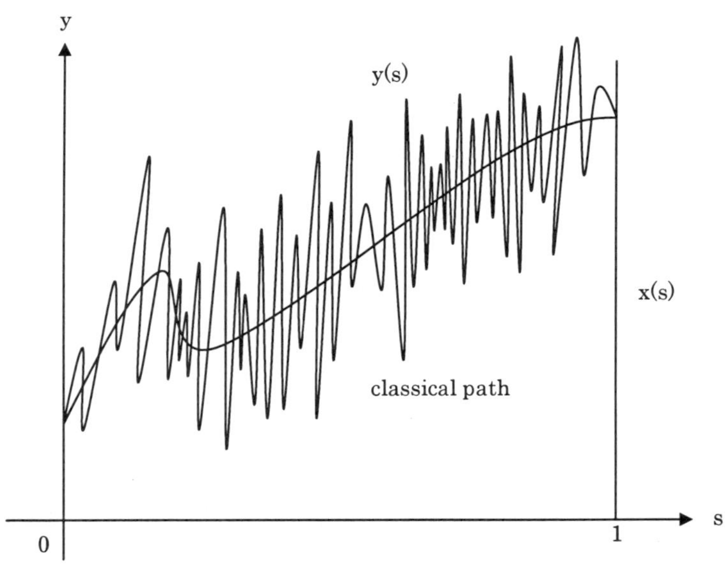

Fig. 10.1.

Now we set

$$S(t_0, t_1) = \int_{t_0}^{t_1} L(t) dt \tag{10.1.6}$$

and

$$\exp\left[\frac{i}{\hbar}\int_{t_0}^{t_1} L(t)dt\right] = \exp\left[\frac{i}{\hbar}S(t_0,t_1)\right] = B(t_0,t_1).$$

Then, we have (see Dirac[21,22]), for $0 < t_1 < t_2 < \cdots < t_n < t$,

$$B(0,t) = B(0,t_1)\cdot B(t_1,t_2)\cdots B(t_n,t).$$

This may be compared with the formula arising from the Markovian semigroup.

Theorem 10.2 *The quantum mechanical propagator $G(0,t;y_1,y_2)$ is given by the following average*

$$G(0,t;y_1,y_2) = \left\langle N\exp\left[\frac{i}{\hbar}\int_0^t L(y,\dot{y})ds + \frac{1}{2}\int_0^t \dot{B}(s)^2 ds\right]\delta_o(y(t)-y_2)\right\rangle. \tag{10.1.7}$$

The average $\langle\ \rangle$ means the integral with respect to the white noise measure μ.

Actual computations for given potentials (including those which have some singularity at the boundary) have been obtained.

We should note that there are generalized white noise functionals involved in the above expectation. Namely, they are delta functions, in fact the Donsker's delta function $\delta_o(y(t)-y_2)$, and a Gauss kernel; the former implicitly appears in the action and the latter is $\exp[\frac{1}{2}\int_0^t \dot{B}(s)^2 ds]$, which is not acceptable by itself because of the constant $\frac{1}{2}$ which is the constant to be excluded. However, we combine it with the term that comes from the kinetic energy $\frac{1}{2}m(\dot{x}(s)+\sqrt{\frac{\hbar}{m}}\dot{B}(s))^2$. The factor $\exp[\frac{1}{2}\int_0^t \dot{B}(s)^2 ds]$ serves as the flattening effect of the white noise measure. One may ask why the functional does so. An intuitive answer to this question is as follows: If we write a Lebesgue measure (exists only virtually) on E^* by dL, the white noise measure μ may be expressed in the form $\exp[-\frac{1}{2}\int_0^t \dot{B}(s)^2 ds]dL$. Hence, the factor in question is put to make the measure μ to be a flat measure dL. In fact, this makes sense eventually.

Returning to the average (10.1.7), which is an integral with respect to the white noise measure μ, it is important to note that the integrand (i.e. the inside of the angular bracket) is integrable, in other words, it is a bilinear form of a generalized functional and a test functional.

In general, the integrand is expressed as a product of a test functional and a functional of the form $\varphi(x) \cdot \delta(\langle x, f\rangle - a)$, where $f \in L^2(R), a \in R$. To this end, we prepare some interpretations.

The first is a short note to be noted. The canonical bilinear form $\langle x, \xi\rangle$ is defined for $\xi \in E$ and $x \in E^*$ (pointwise in x and ξ). For our purpose it is necessary to extend the bilinear form to an extended form $\langle x, f\rangle$, where f is in $L^2(R)$, which we have already introduced in Section 3.1. We call such an extended bilinear form, *a stochastic bilinear form* if it is necessary to discriminate from the ordinary canonical bilinear form.

The note explained below is rather crucial.

The formula (10.1.7) involves a product of functionals of the form like $\varphi(x) \cdot \delta(\langle x, f\rangle - a), f \in L^2(R), a \in \mathbf{C}$. To give a correct interpretation to the expectation of (10.1.7) with this choice, it should be checked that it can be regarded as a bilinear form of a pair of a test functional and a generalized functional. The following assertion answers this question.

Theorem 10.3 *(Streit[91] et al.) Let $\varphi(x)$ be a generalized white noise functional. Assume that the $\mathcal{T}$-transform $(\mathcal{T}\varphi)(\xi), \xi \in E$, of φ is extended to a functional of f in $L^2(R)$, in particular a function of $\xi + \lambda f$, and that $(\mathcal{T}\varphi)(\xi - \lambda f)$ is an integrable function of λ for any fixed ξ. If the transform of $(\mathcal{T}\varphi)(\xi - \lambda f)$ is a U-functional, then the pointwise product $\varphi(x) \cdot \delta(\langle x, f\rangle - a)$ is defined and is a generalized white noise functional.*

Proof. First a formula for the δ-function is provided.

$$\delta_a(t) = \delta(t - a) = \frac{1}{2\pi}\int e^{ia\lambda}e^{-i\lambda x}d\lambda \text{ (in distribution sense).}$$

Hence, for $\varphi \in (S)^*$ and $f \in L^2(R)$ we have

$$\begin{aligned} \mathcal{T}(\varphi(x)\delta(\langle x, f\rangle - a))(\xi) &= \frac{1}{2\pi}\int\int e^{ia\lambda}e^{-i\lambda\langle x, f\rangle}e^{i\langle x,\xi\rangle}\varphi(x)d\mu(x)d\lambda \\ &= \frac{1}{2\pi}\int e^{ia\lambda}(\mathcal{T}\varphi)(\xi^\lambda f)d\lambda. \end{aligned} \quad (10.1.8)$$

By assumption this determines a U-functional, which means the product $\varphi(x)\cdot\delta(\langle x, f\rangle - a)$ makes sense and it is a generalized white noise functional.

■

Example 10.1 The above theorem can be applied to a Gauss kernel $\varphi_c(x) = N \exp[c \int x(t)^2 dt]$.

i) The case where c is real and $c < 0$.
We have

$$(\mathcal{T}\varphi)(\xi - \lambda f) = \exp[\frac{c}{1-2c} \int (\xi(t) - \lambda f(t))^2 dt]$$
$$= \exp[\frac{c}{1-2c}(\|\xi\|^2 - 2\lambda(\xi, f) + \lambda^2 \|f\|^2).$$

This is an integrable function of real λ. Hence, by Theorem 10.3, we have a generalized white noise functional.

Note that $(\mathcal{T}\varphi)(\xi)$ is a characteristic functional, so that it defines a probability measure on E^*.

ii) The case where $c = \frac{1}{2} + ia$, $a \in R - \{0\}$.
The same expression as in i) is given, and it is shown that Theorem 10.3 is applied. Note that this functional appears, in general, in the formula shown in Theorem 10.2.

Example 10.2 In the following case, exact values of the propagators can be obtained (see the literature[160]) and, of course, they agree with the known results.

i) Free particle
ii) Harmonic oscillator.
iii) Potentials which are Fourier transforms of measures.

Recent developments. Now we would like to mention that there are many successful computations of various propagators. We have discussed that in the cases i) free particle, ii) simple harmonic oscillator, iii) the Albeverio-Hohkron potential which is the Fourier transform a measure, the results obtained by our method are in agreement with the known propagators, respectively. In addition, some more interesting cases, including those with much singular potentials and time-dependent potentials, we have satisfactory results in the recent developments as were seen in Theorem 10.3 and Example 10.1. Now follow some more significant examples.

Example 10.3 Streit[91] et al. have obtained explicit formulae in the following cases:

1) a time-dependent Lagrangian of the form

$$L(x(t), \dot{x}(t), t) = \frac{1}{2}m(t)\dot{x}(t)^2 - k(t)^2 x(t)^2 - \dot{f}(t)x(t),$$

where $m(t), k(t)$ and $f(t)$ are smooth functions.

2) A singular potential $V(x)$ of the form

$$V(x) = \sum_n c^{-n^2}\delta_n(x), \quad c > 0,$$

and others.

Our method of path integrals enables us to deal with the case of very irregular potentials to have the propagator. Such a potential cannot be dealt with any other methods, from our best knowledge.

It is significant to see the results by C. C. Bernido and M. V. Carpio Bernido[17]. They are using our method of path integral to investigate the entanglement probabilities of two chain like macro-molecules where one polymer lies on a plane and the other perpendicular to it. The entanglement probabilities are calculated and the result shows a characteristic of the polymer as in the following example.

Example 10.4

Polymer entanglement can be discussed by the white noise approach.

The idea behind the assertion stated above can be extended to other interesting cases, for instance, Chern-Simons' path integral. This fact will be discussed in Section 10.2.

10.2 Hamiltonian dynamics and Chern-Simons functional integral

In recent years there have been increasing interests in the relationship between white noise theory and quantum dynamics in ideal and actual computations. This section is devoted to two topics: 1) symplectic structure in connection with Hamiltonian dynamics and 2) the Chern-Simons action integrals.

1) As soon as we come to Hamiltonian dynamics from Lagrangian formalism with which we are familiar, we are led to noncommutative analysis.

We follow H. Araki's approach [15,16]. Let ∂_t and ∂_t^* be the annihilation and creation operators, respectively. Now t is fixed to make the matters to be one-dimensional. Take a C^2-vector $h = (\alpha, \beta)$ and define an operator

$$B(h) = \alpha\partial_t^* + \beta\partial_t.$$

Define Γ by

$$\Gamma(\alpha, \beta) = (\bar{\beta}, \bar{\alpha}).$$

Set $B(h)^{(*)} = B(\Gamma h)$. Then, for $h_i(\alpha_i, \beta_i), i = 1, 2$, we have

$$[B(h_1)^*, B(h_2)] = i(\bar{\alpha_1}\alpha_2 - \bar{\beta}_1\beta_2).$$

Thus, we can see that there arises a *symplectic* group structure. With this in mind, actually infinite dimensional version of what we have observed, we shall come to differential geometric structure in line with the Hamiltonian dynamics. It is our hope that more intimate connection with white noise theory will be found in this direction.

2) Under the same idea that has been employed in the last section, we can discuss the Chern-Simons-Witten action integral. It is a new application of white noise analysis in line with the approach to path integral by using generalized white noise functionals. There we shall use a generalization of the two dimensional valued of white noise measure introduced in a space of Lie algebra.

What we shall discuss is far from a general theory of Chern-Simons' theory, however we take a basic and rather simpler case following the idea of Albeverio-Sengupta[9] to apply our idea.

There, the authors of the paper[8] propose a 3-dimensional gauge theory based on the Chern-Simons action $CS(A)$. Let their setup be reviewed as follows.

Let G be a compact connected matrix group, and let $\mathbf{g}$ be the Lie algebra associated to the group G. Following the 3-dimensional gauge theory, a connection A is a smooth $\mathbf{g}$-valued 1-form over R^3. It can be expressed in the following form

$$A = a_0 dx_0 + a_1 dx_1 + a_2 dx_2,$$

where a_i's are smooth $\mathbf{g}$-valued functions on R^3. The collection of such A's is denoted by $\mathbf{A}$.

The Chern-Simons action functional $CS(A)$ is given by

$$CS(A) = \frac{\kappa}{4\pi} Tr \left\{ \sum_{i,j,k} \epsilon^{ijk} \int_{R^3} (a_i \partial_j a_k + \frac{2}{3} a_i [A_j, a_k]) dx_0 dx_1 dx_2 \right\}.$$

We assume some conditions on A_i's so that the integral converges. A formal expression of functional integral in question may be expressed in the form

$$\int_A e^{iCS(A)} \varphi(A) \mathcal{D}A,$$

where $\mathcal{D}A$ is an ideal probability measure that we are going to introduce in a mathematically rigorous manner on $\mathbf{A}$.

As was mentioned before we do not intend to discuss very general cases, we assume, at present, that A is gauge invariant to come to a particular case where A is of the form

$$A = \sum_{i=0}^{2} a_i dx_i,$$

with the condition

$$a_2 = 0, \; a_1(x_0, x_1, 0) = 0, \; a_2(x_0, 0, 0) = 0.$$

With these assumptions the $CS(A)$ becomes a simpler formula and can be written as

$$CS(a_0 dx_0 + a_1 dx_1) = \frac{\kappa}{2\pi} Tr \int a_0 \partial_2 a_1 dx_0 dx_1 dx_2.$$

Finally, the Chern-Simons functional integral can be expressed in the form

$$\frac{1}{N} \int_E e^{i\frac{\kappa}{2\pi}} \langle a_0, b_1 \rangle \mathcal{D}a_0 \mathcal{D}b_1,$$

where E is a space of all pairs (a_0, b_1), N is a normalizing constant and $b_1 = \partial_2 a_1$.

As is promised before, we are now ready to introduce an expression of $\mathcal{D}a_0\mathcal{D}b_1$ in terms of two-dimensional valued white noise measure ν. It is given by $\nu = \mu_x \times \mu_y$ on $E_x^* \times E_y^*$ which is the product space of two-dimensional valued generalized functions (a_0, b_1).

Now remains a serious problem. Namely, we have to check the integrability of the Chern-Simons functional integral even in such a simple and particular case.

Observe the factor $e^{ic\langle a_0, b_1\rangle} = e^{ic\langle x,y\rangle}$ with positive constant c. To carry on the integral with respect to a suitable measure on the space $E^*_{(x,y)}$, the following arguments may give some suggestion to proceed further investigation.

Assume that the collection of (x, y)'s is the product space $E_x^* \times E_y^*$ of the spaces E_x^* and E_y^* of generalized functions, respectively. The product measure $\nu(dz) = \mu_X(dx) \times \mu_Y(dy)$, with $dz = dx \wedge dy$, is introduced to $E_x^* \times E_y^*$.

It is obvious that

Proposition 10.1 *The probability measure $\nu(dxdy)$ is invariant under the rotations acting on (x, y)-space.*

In particular, under the transformation

$$z = (x, y) \;\rightarrow\; (x + y, x - y)/\sqrt{2}$$

the measure $\nu(dz)$ is invariant.

We note an obvious equality

$$\langle x, y\rangle = \frac{1}{2}(\|x + y\|^2 - \|x - y\|^2).$$

Although the equality is rather formal, it is guaranteed within the framework of generalized white noise functionals. Also, we note that $(x + y)$ and $(x - y)$ are independent with respect to the Gaussian measure ν. Hence, the factor $e^{ic\langle x,y\rangle}$, with c constant, can be reduced to the product of independent Gauss kernels that are of the form

$$N \exp[c \int : x(t)^2 : dt],$$

which is familiar for us (see Example 2.5 in Section 2.6).

Now the integral in question can be well defined with the choice φ to be a test functional of a connection. More precisely, first take a degenerated functional written as a product of functionals of $x + y$ and $x - y$. The functional integration is obvious. Then, go to the general test functionals.

There are various stimulative literatures in this direction by many authors; for instance, S. Albeverio, A. Hahn and A. Sengupta[8] ; S. Albeverio, and A. Sengupta[9]; A. Hahn[34]; R. Léandre; P. Leukert and J. Schöfer[98]; M.B. Green, J.H. Schwarz and E. Witten[32], II, and so forth. It is our hope that interesting developments will be obtained in line with white noise analysis.

Note We also mention an interest stimulating result by Léandre[95] that deals with connections between geometry and white noise analysis.

Further it is possible to discuss an application to non-commutative geometry through our path integral method. Good references are Léandre[95,96], etc.

10.3 Dirichlet forms

There are many ways to construct Markov processes under various and different conditions. Among others, the method of using Dirichlet form is interesting not only by itself, but also noted by the beautiful connection with other fields of analysis, in particular potential theory.

It is also mentioned that the method of Dirichlet forms gives us concrete tools to define strong Markov processes. It is known that Dirichlet form determines a symmetric Markov semigroup on L^2-space. Assuming, in addition, the regularity we are given a Markov process with *strong* Markov property.

To discuss Dirichlet forms as an application of white noise analysis, it is necessary to prepare some facts on positive generalized white noise functionals.

A functional φ in (S) is called *positive* if

$$\varphi(x) \geq 0, \quad a.e.\ \mu.$$

As for a generalized functional $\psi(x)$ we define in the following manner. If ψ satisfies

$$\langle \psi\ \varphi \rangle \geq 0, \quad \text{for every positive } \varphi,$$

then, ψ is said to be *positive.*

The collection of all positive generalized functionals is denoted by $(S)^*_+$ which forms a convex cone. It is known (Kubo-Yokoi[90]) that any φ in (S) has a continuous version, denoted by $\tilde{\varphi}(x)$. With this result, we claim that for any $\psi \in (S)^*_+$ there exists a measure ν_ψ such that

$$\langle \psi, \varphi \rangle = \int \tilde{\varphi}(x) d\nu_\psi. \tag{10.3.1}$$

The following assertion is due to Y. Yokoi[169].

Proposition 10.2 *The $\mathcal{T}$-transform of a positive generalized white noise functional is positive definite. The converse is true.*

Example 10.5 Again consider the Gauss kernel

$$\varphi_c(x) = N \exp[c \int x(t)^2 dt],$$

where the constant c is chosen such that $c < \frac{1}{2}$ holds, and where N is the renormalizing constant 2.

The $\mathcal{T}$-transform of φ_c is $\exp[\frac{1}{2(1-2c)} \int \xi(u)^2 du]$, which is positive definite. Hence

$$\varphi_c \in (S)^*_+.$$

Example 10.6 Consider a quadratic functional of x;

$$\psi(x) = \int_0^1 : x(t)^2 : dt.$$

The S-transform of $\psi(x)$ is

$$U(\xi) = \int_0^1 \xi(t)^2 dt.$$

It is proved that $\psi(x)$ is *not* positive generalized functional, quite contrary to what we expect.

Actually, a proof of this fact is given as follows.

Take a white noise functional $\varphi(x)$ defined by

$$\varphi(x) = \exp[-\langle x, \eta\rangle^2],$$

where $\eta \in E$ and $\|\eta\| = 1$. Obviously φ is positive and is a test functional. Now the value of $\psi(x)$ is evaluated at φ. For this purpose the S-transform of φ is computed.

$$(S\varphi)(\xi) = C(\xi) \int \exp[\langle x, \xi\rangle] \cdot \exp[-\langle x, \eta\rangle^2] d\mu(x).$$

Set $\xi = a\eta + \eta'$ with $a = \langle \xi, \eta\rangle$ and $\eta \perp \eta'$. Then,

$$(S\varphi)(\xi) = C(\xi) \int \exp[a\langle x, \eta\rangle - \langle x, \eta\rangle^2] d\mu(x) \int \exp[\langle x, \eta'\rangle] d\mu(x),$$

since $\langle x, \eta\rangle$ and $\langle x, \eta'\rangle$ are independent. By elementary computations the above integral is proved to be equal to

$$3^{-1/2} \exp[-\frac{1}{3}(\xi, \eta)^2].$$

The S-transform of the H_2-component (the term of degree 2) of the chaos expansion of φ is of the form $3^{-3/2}(\xi, \eta)^2$. The kernel of its integral representation is therefore $-3^{-3/2}\eta^{2\otimes}(u, v)$. While the kernel of ψ is $\delta(u - v), (u, v) \in [0, 1]^2$. Hence, the generalized functional ψ evaluated at the test functional φ is $-2 \cdot 3^{-3/2}(\xi, \eta)^2$, which is negative. Thus, the assertion is proved.

Energy forms and Dirichlet forms.

Recall the operator ∂_t, which is an annihilation operator acting on white noise functionals. Intuitively, it means an operator that may be expressed as $\frac{\partial}{\partial \dot{B}(t)}$. Its domain is a subset of the space $(L^2)^-$. Now, the ∇ operator is applied to have $\nabla\varphi$:

$$\nabla\varphi = (\partial_t \varphi), \quad t \in R^1, \quad \varphi \in (S). \tag{10.3.2}$$

More precisely, set

$$(\mathcal{L}^2) = (L^2) \otimes L^2(R^1).$$

Then, the operator ∇ defines a continuous mapping

$$\nabla : \ (L^2) \longrightarrow \ (\mathcal{L}^2),$$

determined by $\nabla\varphi = (\partial_t \varphi)$ for $\varphi \in (S)$ and for $t \in R$.

Also, we set

$$|\nabla\varphi|^2 = \int_R |\partial_t \varphi|^2 dt.$$

Then, we have

Proposition 10.3 *If φ is in (S), then*

$$|\nabla\varphi|^2 \in (S).$$

This is proved by evaluating the p-th norm $\| |\nabla\varphi|^2 \|_p$, where we use the fact that the (S) is an algebra. (See Proposition 2.5.)

We are now ready to use the notion "positive generalized white noise functionals" discussed above. With such positive generalized functionals *energy form* can be defined as follows.

Let Φ be a positive generalized white noise functional. Then, there is a unique positive finite measure ν on the measurable space $(E^*, \mathbf{B})$ such that for $\varphi \in (S)$

$$\langle \Phi, \varphi \rangle = \int \tilde{\varphi} d\nu,$$

where $\tilde{\varphi}$ is the continuous version of φ. (See Ito-Kubo[78].)

Remark 10.1 *Often $d\nu$ is written as $\Phi d\mu$, however this does not mean that $d\nu$ is absolutely continuous with respect to $d\mu$. Note that Φ is a generalized functional.*

Now energy form $\mathcal{E}(\varphi)$ is defined by

$$\mathcal{E}(\varphi) = \int |\nabla\varphi|^2 d\nu.$$

Theorem 10.4 *(Fukushima[27]) The energy form $\mathcal{E}$ defined by a positive generalized white noise functional Φ is a positive, densely defined, symmetric quadratic form on $L^2(\nu)$.*

Proof is given by noting that the domain $\mathcal{D}(\mathcal{E}) = (S)$ is dense in $L^2(\nu)$ and other obvious facts.

Definition 10.1 A positive generalized white noise functional Φ is called *admissible* if the corresponding energy form is closable on $L^2(\nu)$.

Denote by ∇^* the adjoint of the operator ∇. Our main result is

Theorem 10.5 *If*

$$\nabla^* : \mathcal{L}^2(\nu) \longrightarrow (L^2(\nu))$$

is densely defined, then $\mathcal{E}$ is admissible.

The following theorem is useful in actual applications.

Theorem 10.6 *A positive generalized functional Φ is admissible if there is a functional B_t satisfying*

$$\partial_t \Phi = B_t \Phi,$$

for every $t \in R$.

Example 10.7 If K maps (S) into itself, then we see that the following functional Φ is admissible.

$$\Phi(x) = N \exp[-\frac{1}{2}\langle x, Kx\rangle],$$

where N is the normalizing constant. We note that

$$B_t(x) = -\langle x, K(t, \cdot)\rangle.$$

Theorem 10.7 *Let Φ be an admissible positive generalized functional and let $\bar{\mathcal{E}}$ be the closure of the associated energy form $\mathcal{E}$. Then, $\mathcal{E}$ is Markovian in the sense of Fukushima[27].*

A proof is given in a similar manner to the finite dimensional case, so it is omitted.

10.4 Time operator

The theory of time operator has been investigated since many years ago within the study of random evolutional systems, where the time development is described by a one-parameter group of unitary operators parametrized by the time variable t.

There are two ways of developing the theory of time operator. One is a probabilistic approach. The basic idea is to have a representation of the time t by a stationary stochastic process $X(t)$, where $R = \{t\}$ is understood as an additive group and at the same time it is considered as a linearly ordered continuum. The second is a quantum mechanical approach. The time is viewed as a self-adjoint operator, denoted by T, which should satisfy the commutation relations of Heisenberg type with the Hamiltonian H.

Both ways are significant in their own purpose. Our present aim is to investigate representation of time variable t in terms of certain stationary stochastic process. This is, in reality, natural to discuss in connection with the white noise analysis.

Our game is therefore played on a measure space or Hilbert space constructed by the stationary stochastic process.

We know that, in the study of a flow (or a dynamical system) on a suitable measure space, some additional one-parameter group of operators satisfying a certain relation with the given flow plays important roles. The relationship should be expressed in terms of a commutation relation between the associated infinitesimal generators.

A general algebraic theory of commutation relations between two members may be stated as follows. Let α and β together with the identity I form a complex Lie algebra under the Lie product $[\cdot, \cdot]$. There three typical cases for a pair α and β are possible :

1) commutative: $[\alpha, \beta] = 0$,
2) one is transversal to the other: $[\alpha, \beta] = \alpha$,
3) canonical commutation relation: $[\alpha, \beta] = iI$.

The algebraic structure of each of them are as follows.

1) is trivial. For 2), form a Lie algebra **a** generated by α and β. Then **a** is solvable. For 3), the algebra generated by $\{\alpha, \beta, I\}$ is solvable and nilpotent. Those members indicate a set of standard possible algebraic structures. To be very interesting, they, except the trivial case, appear in the framework of white noise analysis dealing with stationary stochastic processes, as is seen in what follows.

We are now in a position to setup the problems on time operator. To this end, it is necessary to prepare some background on stochastic process. It seems better to start with elementary and in fact classical examples rather than abstract approaches.

Let $X(t) = X(t, \omega)$, $t \in R$, $\omega \in \Omega(P)$, be a stationary second order stochastic process. ($X(t)$ is continuous in mean square and $E(X(t)) = 0$). Assume that $X(t)$ is purely non-deterministic. Define $M_t(X)$ to be a closed subspace of $L^2 = L^2(\Omega, P)$ spanned by $X(s), s \leq t$. Then $M_t(X)$ is an increasing sequence of subspaces, and

$$\bigvee_{t \in R} M_t(X)$$

is denoted by $M(X)$ which is a Hilbert space ($\subset L^2$). The assumption that $X(t)$ is purely nondeterministic is now expressed as

$$\bigcap_{t \in R} M_t(X) = \{0\}.$$

With this background, a projection operator $E(t)$ is defined in such a manner that :

$$E(t) : M(X) \mapsto M_t(X).$$

Continuity of $X(t)$ implies that of $E(t)$ in t. The collection $\{E(t); t \in R\}$ turns out to be a resolution of the identity I on the Hilbert space $M(X)$. Namely, we have, by definition

$$E(+\infty) = I,$$
$$E(t)E(s) = E(t \wedge s),$$

and by assumption

$$E(-\infty) = 0.$$

Definition 10.2 A self-adjoint operator T on the space $M(X)$ given by

$$T = \int t dE(t)$$

is well-defined and is called the *time operator.*

The domain $D(T)$ of T is dense in $M(X)$.

On the other hand, there is a mapping U_t such that

$$U_t : \ U_t X(s) = X(s+t).$$

It is an isometry, so that it extends, by linearity, to a unitary operator acting on the Hilbert space $M(X)$; still denoted by the same notation U_t. By definition it is proved that

$$U_t U_s = U_{t+s}.$$

By the mean continuity of $X(t)$, we can prove that

$$U_t \ \rightarrow \ I, \text{ as } \ t \rightarrow \ 0.$$

Thus, we are given a continuous one-parameter group $U_t, t \in R$, of unitary operators on $M(X)$.

Now one may be interested in the commutation relations between the time operator and the unitary group U_t that is familiar for us to study stationary processes. It is easy to show the following (we refer to the paper[13]).

Proposition 10.4 *It holds that*

$$TU_t = U_t T + tU_t. \tag{10.4.1}$$

Recall Stone's Theorem for continuous one-parameter unitary group. Actually, we have a self-adjoint operator H such that

$$U_t = \exp[itH].$$

Hence,

Corollary 10.1 *The commutation relation*

$$[H, T] = iI$$

holds.

This relation is useful in the study of stationary stochastic process $X(t)$. The technique is the same as in the case of quantum mechanics where the Heisenberg uncertainty principle plays a dominant role. Thus, we can learn much from quantum dynamics.

If, in particular, the $X(t), t \in R$, is Gaussian, one can find interesting connection with the *canonical representation* theory of $X(t)$. Namely, if there exists a canonical representation of $X(t)$, and if it is 1-ple Markov, then the canonical representation is expressed in the form

$$X(t) = f(t) \int_{-\infty}^{t} g(u) dB(u) = f(t) U(t),$$

where $B(u)$ is an additive process and where $f(t)$ never vanishes.

Now assume that $g(u)^2 E(|dB(u)|^2)$ is a measure equivalent to Lebesgue measure. Let the density function be denoted by $p(u)$, which is to be positive almost everywhere. Then, set $dB_0(u) = p(u)^{-1/2} dB(u)$. It is easy to see that $dB_0(u)$ is a variation of a Brownian motion, so that it is a standard Gaussian random measure. We can also see that $M_t(B_0) = M_t(X)$. Then we can form a stationary, simple Markov Gaussian process $Y(t)$ such that

$$M_t(X) = M_t(Y)$$

for every t. This fact suggests us that we can find a time operator T^X that describes the propagation of the random events determined by the $X(t)$, so that we can play the same game as those for a purely non-deterministic weakly stationary process.

If we are given a *multiple Markov* (say, N-ple Markov) stationary Gaussian process which is purely nondeterministic, then the canonical representation exists and can be expressed in the form of a Goursat kernel (see Hida[35]):

$$X(t) = \sum_{1}^{N} f_i(t)U_i(t), \tag{10.4.2}$$

where $\{f_i(t)\}$ is a fundamental system of solutions of ordinary differential equation with constant coefficients, and where $U_i(t)$'s are additive Gaussian processes such that $\{U_i(t)\}$ are linearly independent system for every t. In addition, we can prove that a vector-valued process

$$\mathbf{U}(t) = (U_1(t), U_2(t), \cdots, U_N(t))$$

is additive, hence it is Markov. The above discussion can, therefore, be extended to the case of a simple Markov vector-valued Gaussian process expressed in the form $\mathbf{f}(t)\mathbf{U}(t)^*$, $\mathbf{f}$ and $\mathbf{U}$ being vectors.

There is a multiple Markov Gaussian process that can be reduced to the case of a weakly *stationary* process. Suppose an N-ple Markov Gaussian process $X(t), t \geq 0$, is *dilation quasi-invariant*, namely $X(at)$ for $a > 0$, is the same process as $a^{1/2}X(t)$. Then, by changing the time parameter t to e^t, we are given a stationary N-ple Markov process. So the above trick can be applied.

A *nonlinear* version of the time operator theory can also be discussed. The basic Hilbert space is now taken to be the space involving all square integrable functions which are measurable with respect to the sigma-fields generated by the given stationary process $X(t)$. In this case, we do not need to assume the existence of moment of any order of the stationary process $X(t)$.

Let $\mathbf{B}_t(X)$ be the sigma-field generated by measurable subsets (of Ω) determined by the $X(s,\omega), s \leq t, \omega \in \Omega$. In other words, $\mathbf{B}_t(X)$ is the smallest sigma-field with respect to which all the $X(s), s \leq t$, are measurable. Set $\mathbf{B}(X) = \bigvee_t \mathbf{B}_t(X)$. Then, we have Hilbert spaces $L_t^2(X) = L^2(\Omega, \mathbf{B}_t(X), P)$ and $L^2(X) = L^2(\Omega, \mathbf{B}(X), P)$, respectively. There is naturally defined an orthogonal projection $E'(t) \ : \ L^2(X) \to L_t^2(X), \ t \in R$.

Set $E(t) = E'(t+)$. The collection $\{E(t); \ t \in R\}$ forms a resolution of the identity on the Hilbert space $L^2(X)$. The operator T is given by the same expression as in the case of $M(X)$,

$$T = \int t dE(t)$$

and is called the *time operator* on $L^2(X)$.

On the other hand, we can define a *unitary group of shift operators* $\{U_t,\ t \in R\}$. Starting with a mapping V_t :

$$V_t X(s) = X(t+s).$$

Then, V_t extends to a unitary operator on $L^2(X)$. Since V_t is, by definition, multiplicative, it defines a *flow* (one-parameter group of measure preserving transformations) $\{T_t, t \in R\}$ on the measure space $(\omega, \mathbf{B}(X), P)$ such that

$$T_t : \mathbf{B}_s(X) \quad \mapsto \quad \mathbf{B}_{s+t}(X).$$

More precisely, for a cylinder set, say an ω-set :

$$A = \{\omega : (X(s_1, \omega), \cdots, X(s_n, \omega)) \in B_k\}$$

being a Borel subset in R^n, we define $T_t A$ by

$$T_t A = \{\omega : (V_t X(s_1, \omega), \cdots, V_t X(s_n, \omega)) \in B_k\}.$$

Obviously, T_t is a measurable transformation and

$$P(T_t A) = P(A)$$

holds. Then, the T_t extends to a measure preserving transformation on the algebra $\mathbf{A}(X)$ generated by all the cylinder sets of the form A above, where the operations union and intersection applied to cylinder subsets commute with T_t. Since the sigma-field $\mathbf{B}(X)$ is generated by $\mathbf{A}(X)$, T_t can be extended to a measure preserving transformation on $(\Omega, \mathbf{B}(X), P)$. It is easy to prove the group property

$$T_t T_s = T_{t+s},$$

namely $\{T_t, t \in R\}$ forms a one-parameter group with $T_0 = I$.

To proceed to the next step we assume that

Assumption The measure space $(\Omega, \mathbf{B}(X), P)$ that we shall deal with is an abstract Lebesgue space.

More concretely, the measure space is isomorphic to a sum of Lebesgue measure space on a finite interval and point measures (atoms).

It is known that for an abstract Lebesgue space there is a countable basis.

We can say that the measure space $(\Omega, \mathbf{B}(X), P)$, on which the stationary process $X(t)$ is defined, is isomorphic to an abstract Lebesgue space without atoms.

Example 10.8

1) If $X(t)$ is a continuous stationary Gaussian process with canonical representation, then the space $(\Omega, \mathbf{B}(X), P)$ is an abstract Lebesgue space without atoms.

2) For a linear process expressed as a linear functional of a white noise and a Poisson noise, the associated measure space is also an abstract Lebesgue space without atoms.

Assuming that the given measure space is an abstract Lebesgue space, the family of set transformations $\{T_t\}$ turns into a family of point transformations. This fact can be proved with the help of countable basis.

In addition, we can see that the family $\{T_t\}$ forms a one-parameter group of measure preserving transformations on the measure space $(\Omega, \mathbf{B}(X), P)$ except for a null set. Also, by the assumption of the continuity of $X(t)$, it can be proved that

$$T_t\omega \text{ is measurable in } (t, \omega).$$

Hence, by the usual argument we can prove the following assertion.

Proposition 10.5 *Set*

$$(U_t\varphi)(x) = \varphi(T_t x).$$

Then, $\{U_t, t \in R\}$ forms a continuous one-parameter group of unitary operators acting on the Hilbert space $L^2(X)$:

$$\begin{aligned} U_t U_s &= U_{t+s}, \\ U_t &\to I \quad \textit{as } t \to 0. \end{aligned}$$

Having obtained the unitary group $\{U_t\}$, we can prove the commutation relation with the time operator T established before. Note that the commutation relation is exactly the same in expression as in the linear case where the entire Hilbert space is taken to be $M(X)$.

Theorem 10.8 *It holds that*

$$TU_t = U_t T + tU_t.$$

Apply Stone's theorem to U_t to have the infinitesimal generator which is self-adjoint and is denoted by H :

$$\frac{d}{dt}U_t\Big|_{t=0} = H.$$

Then, we have the commutation relation

$$[H, T] = iI.$$

Thus, similar results are obtained as in $M(X)$, however, there is a short note to be added. Since $M(X) \subset L^2(X)$ holds and two spaces have been introduced consistently so far as $X(t)$ has second order moment, all the statements commute with the projection operator $P : L^2(X) \to M(X)$.

It is hoped that the above relationships among operators will give some help to the profound investigation of the actual evolutional phenomena described by $X(t)$.

We now turn our eyes to a stationary *Markov* process $X(t)$ with $t \geq 0$. Assume that the transition probability density $p(t, x, y)$ exists and is smooth in (t, u, v). Define an operator V_t, $t \geq 0$, by

$$V_t f(u) = \int p(t, u, v) f(v) dv, \quad f \in \mathbf{C}.$$

Then, we have a semigroup of operators V_t acting on $\mathbf{C}$:

$$\begin{aligned} V_t V_s &= V_{t+s}, \\ V_t &\to I, \quad (t \to 0). \end{aligned}$$

We can therefore appeal to the Yosida-Hille theorem for one-parameter semigroup to obtain a generator A such that :

$$w - \lim_{t\to 0} \frac{V_t - I}{t} f = Af$$

for f in a dense subset of $\mathbf{C}$.

Example 10.9 We consider the case where $X(t)$ satisfies the *Langevin equation*:

$$dX(t) = -\lambda X(t)dt + dB(t).$$

In the white noise terminology it is written as

$$\frac{d}{dt}X(t) = \lambda X(t) + \dot{B}(t).$$

Then, A has an explicit expression of the form

$$A = \frac{1}{2}\frac{d^2}{du^2} - \lambda u \frac{d}{du}.$$

With the operator A given above, we can establish an interesting commutation relation. Namely, simple computation implies

Proposition 10.6 *Let A be as in Example 10.9 and let $S = \frac{d}{du}$. Then, we have*

$$[A, S] = \lambda S.$$

Thus, a *transversal relation* appears. This is a typical relationship in the theory of dynamical systems, in particular in the case where ergodic property is investigated (cf. Ya. G. Sinai[157] and I. Kubo[88]). Actually, this fact tells us a basic property from the viewpoint of a dynamical system that is determined by the Langevin equation.

In terms of dynamical systems, one may say that the shift is *transversal* to V_t. From this viewpoint profound roles of the time operator are found (see the monograph[71], Chapter 10).

In addition, the operator S is the generator of the shift (the one-parameter group of translations of space variable), so that we can use harmonic analysis arising from the shift and the semigroup of the phase transition.

10.5 Addendum: Euclidean fields

This topic has been discussed for more than three decades and many results have been obtained successively. In addition, good literatures have been published so far, thus there may be no need to devote one section to this topic. A good connection with the white noise theory has been discovered by Albeverio, Yoshida and others, so this section is added to an announcement of new results on Euclidean field theory.

The Euclidean fields are, roughly speaking, obtained by letting the time t be imaginary, so that the Schrödinger equation becomes diffusion equation. By doing so we can escape from the awkwardness of non-integrability. This direction has been much established and at present good literatures are available. See e.g. J. Glim and A. Jaffe[31], and B. Simon[156] and others. We therefore do not want to go into details on this topic, but it is worth mentioning and even its importance is emphasized.

We have been informed that there are recent developments on this topic with slightly different direction, however those subjects are somewhat far from our present approach.

It should be noted that the constructive field theory in quantum mechanics is an important problem and has been discussed by many authors and is now developing steadily with close connection with the Euclidean field theory.

Related to the time operator discussed in the last section, we make a brief comment on the study of Euclidean fields. The time reflection is defined. It has been shown that the *reflection positivity* guarantees that a Euclidean field with this property can be transformed to a quantum field with real space-time parameter. More details on Euclidean field theory, we refer to the monographs[155,156]. It is also noted that Albeverio and Yoshida has recently obtained an interesting method of constructing a non-Gaussian reflection positive generalized random fields (see the literature[10]). Also the authors have published quite an interesting paper on this topic by using generalized white noise functionals. For detail see Albeverio and Yoshida[11].

Chapter 11

Appendix

Hermite polynomial $H_n(x)$

Definition 11.1 Hermite polynomial of degree n is defined by

$$H_n(x) = (-1)^n e^{x^2} \frac{d^n}{dx^n} e^{-x^2}, \quad n \geq 0.$$

Moment Generating Function

$\sum_{n=1}^{\infty} \frac{t^n}{n!} H_n(x) = e^{-t^2+2tx}$

$H_n(x) = n! \sum_{k=0}^{[\frac{n}{2}]} \frac{(-1)^k}{k!} \frac{(2x)^{n-2k}}{(n-2k)!}$

Examples

$H_0(x) = 1; \quad H_1(x) = 2x; \quad H_2(x) = 4x^2 - 2$

$H_3(x) = 8x^3 - 12x; \quad H_4(x) = 16x^4 - 48x^2; \quad H_5(x) = 32x^5 - 160x^3 + 120x$

$H_n''(x) - 2xH_n'(x) + 2nH_n(x) = 0$

$H_n'(x) = 2nH_{n-1}(x)$

$H_{n+1}(x) - 2xH_n(x) + 2nH_{n-1}(x) = 0$

$H_n(ax + \sqrt{1-a^2}y) = \sum_{k=0}^{n} \binom{n}{k} a^{n-k} (1-a^2)^{k/2} H_{n-1}(x) H_k(y)$

Particular Cases

$$H_n(\frac{x+y}{\sqrt{2}}) = 2^{-n/2}\sum_{k=0}^{n}\binom{n}{k}H_{n-k}(x)H_k(y)$$

$$H_m(x)H_n(x) = \sum_{k=0}^{m\wedge n} 2^k k!\binom{m}{k}\binom{n}{k}H_{m+n-k}(x)$$

$$\int_{-\infty}^{\infty} H_m(x)H_n(x)e^{-x^2}dx = \sqrt{\pi}\delta_{n;m}2^n n!\sqrt{\pi}$$

$$\int_{-\infty}^{\infty} H_m(\frac{x}{\sqrt{2}})H_n(\frac{x}{\sqrt{2}})e^{-x^2/2}dx = \sqrt{2\pi}\delta_{n;m}2^n n!$$

Set $\xi_n(x) = \frac{1}{\sqrt{2^n n!\pi^{1/4}}}H_n(x)e^{-x^2/2}$: $\{\xi_n; n \geq 0\}$, is a complete orthonormal system in $L^2(R)$.

$$\xi_n'(x) = \sqrt{\tfrac{n}{2}}\xi_{n-1}(x) - \sqrt{\tfrac{n+1}{2}}\xi_{n+1}(x)$$

$$x\xi_n(x) = \sqrt{\tfrac{n}{2}}\xi_{n-1}(x) + \sqrt{\tfrac{n+1}{2}}\xi_{n+1}(x)$$

$$(-\tfrac{d^2}{dx^2} + x^2 + 1)\xi_n = (2n+2)\xi_n$$

$$\frac{1}{\sqrt{2\pi}}\int_{-\infty}^{\infty} e^{ixy}H_n(y)e^{-y^2/2}dy = i^n H_n(x)e^{-\frac{x^2}{2}}$$

$$\frac{1}{\sqrt{2\pi}}\int_{-\infty}^{\infty} (x+iy)^n e^{-y^2/2}dy = 2^{-n/2}H_n(\frac{x}{\sqrt{2}})$$

$$\frac{1}{\sqrt{2\pi}}\int_{-\infty}^{\infty} (H_n(\frac{i}{\sqrt{2}}x + y)e^{-y^2/2}dy = i^n H_n(\frac{x}{\sqrt{2}})$$

Hermite polynomials with parameter

Definition 11.2 Hermite polynomial of degree n with parameter is defined as

$$H_n(x;\sigma^2) = \frac{(-\sigma^2)^n}{n!}e^{\frac{x^2}{2\sigma^2}}\frac{d^n}{dx^n}e^{-\frac{x^2}{2\sigma^2}}; \quad \sigma > 0;\ n \geq 0$$

Generating function

$$\sum_{n=1}^{\infty} \frac{t^n}{n!} H_n(x;\sigma^2) = e^{-\frac{\sigma^2}{2}t^2+tx}$$

$$H_n(x;\sigma^2) = \frac{\sigma^{2^n}}{n!2^{n/2}} H_n(\frac{x}{\sqrt{2}\sigma})$$

$$H_n''(x;\sigma^2) - \frac{x}{\sigma^2} H_n'(x;\sigma^2) + \frac{n}{\sigma^2} H_n(x;\sigma^2) = 0$$

$$H_n'(x;\sigma^2) = H_{n-1}(x;\sigma^2)$$

$$H_{n+1}(x;\sigma^2) - \frac{x}{n+1} H_n(x;\sigma^2) + \frac{\sigma^2}{n+1} H_{n-1}(x;\sigma^2) = 0$$

$$\sum_{k=0}^{n} H_{n-k}(x;\sigma^2) H_k(y;\tau^2) = H_n(x+y;\sigma^2+\tau^2)$$

$$H_m(x;\sigma^2) H_n(x;\sigma^2) = \sum_{k=0}^{m\wedge n} \frac{\sigma^{2k}(m+n-2k)!}{k!(m-k)!(n-k)!} H_{m+n-2k}(x;\sigma^2)$$

Set $\eta_n(x;\sigma^2) = \frac{\sqrt{n!}\sigma^n}{H}{}_n(x;\sigma^2)$, $\{\eta_n; n \geq 0\}$ is a complete orthonormal Gaussian system in $L^2(R, \frac{1}{\sqrt{2\pi}\sigma} e^{-\frac{x^2}{2\sigma^2}} dx)$.

Complete Hermite polynomial

Definition $H_{p,q}(z,\bar{z}) = (-1)^{p+q} e^{z\bar{z}} \frac{\partial^{p+q}}{\partial \bar{z}^p \partial z^q} e^{-z\bar{z}}, z \in C, p, q \geq 0$

Generating Function

$$\sum_{p,q=0}^{\infty} \frac{\bar{t}^p t^q}{p!q!} H_{p,q}(z,\bar{z}) = e^{-t\bar{t}+t\bar{z}+\bar{t}z}, \ t \in C$$

$$H_{p,q}(z,\bar{z}) = \sum_{k=0}^{p\wedge q} (-1)^k \frac{p.q!}{k!(p-k)(q-k)!} z^{p-k} \bar{z}^{q-k}$$

$$H_{p,q}(z,\bar{z}) = \overline{H_{q,p}(z,\bar{z})}$$

$\frac{\partial^2}{\partial z \partial \bar{z}} H_{p,q}(z, \bar{z}) - \bar{z} \frac{\partial}{\partial \bar{z}} H_{p,q}(z, \bar{z}) + q H_{p,q}(z, \bar{z}) = 0$

$\frac{\partial^2}{\partial \bar{z} \partial z} H_{p,q}(z, \bar{z}) - z \frac{\partial}{\partial z} H_{p,q}(z, \bar{z}) + q H_{p,q}(z, \bar{z}) = 0$

$\frac{\partial}{\partial z} H_{p,q}(z, \bar{z}) = p H_{p-1,q}(z, \bar{z})$

$\frac{\partial}{\partial \bar{z}} H_{p,q}(z, \bar{z}) = q H_{p-1,q}(z, \bar{z})$

$H_{p+1,q}(z, \bar{z}) - z H_{p,q}(z, \bar{z}) + q H_{p,q-1}(z, \bar{z})$

$H_{p,q+1}(z, \bar{z}) - \bar{z} H_{p,q}(z, \bar{z}) + p H_{p-1,q}(z, \bar{z})$

$\{\frac{1}{\sqrt{p!q!}} H_{p,q}(z, \bar{z}); p \geq 0, q \geq 0\}$ is a complete orthonormal system in

$L^2(C, \frac{i}{2\pi} e^{-z\bar{z}} dz \wedge d\bar{z})$.

$\sum_{p+q=n} \frac{n!}{p!q!} H_{p,q}(x, x) = H_n(x)$, x is real number.

Bibliography

1. L. Accardi, Quantum information, quantum communication and information. Quantum Information and Complexity, ed. T. Hida et al. World Scientific Pub. Co. 2004, 1-60.
2. L. Accardi, White noise calculus and stochastic calculus. Proc. 2004 Nagoya Conf Proc. Part II. World Scientific Pub. Co. to appear.
3. L. Accardi, Y. G. Lu and I. Volovich, Quantum theory and its stochastic limit. Apringer-Verlag. 2002.
4. L. Accardi et al eds. Selected Papers of Takeyuki Hida. World Scientific Pub. Co. Ltd. 2001.
5. L. Accardi, T. Hida and Si Si, Innovation approach to some stochastic processes. Volterra Center Notes. N.537, 2002.
6. L. Accardi, Y.G. Lu and I. Volovich, Quantum theory and its stochastic limit. Springer-Verlag, 2002.
7. L. Accardi, H. Ouerdiane and O.G. Smolyanov, Lévy Laplacian acting on operators. Russian Journal of Mathematical Physics. 10 (2003), 359-380,
8. S. Albeverio, A. Hahn and A. Sengupta, Rigorous Feynman path Integrals, with applications to quantum theory, gauge fields, and topological invariants. Stochastic Analysis and Mathematical Physics. ed. R. Rebolledo et al. World Scientific Pub. Co. 2002, 1-60.
9. S. Albeverio and A. Sengupta, The Chern-Simons functional integral as an infinite dimensional distribution. Nonlinear Analysis, Theory, Methods and Applications. 30 (1997), 329-335.
10. S. Albeverio and M.W. Yoshida, Multiple stochastic integerals construction of non-Gaussian reflection positive generalized random fields. Preprint 2006.
11. S. Albeverio and M W. Yoshida, Markovian semigroups, time translation operator and a new development of Hida product between Hida distributions on sharp time fields, Micro-Macro duality in quantum analysis, RIMS pub. Kyoto Univ. ed. I. Ojima 2007. 28-35..
12. E. Alòs, O.Mazet and D. Nualart, Stochastic calculus with respect to Gaussian processes. Ann. Probability, 29 (2001), 766-801.
13. I. Antoniou and Z. Suchanecki, On the quantum Liouville space. Proc. Quantum Information and Complexity, Nagoya. eds, T. Hida, K. Saitô and

Si Si, 2004, 61-86.
14. H. Araki and M. Shiraishi, On quasifree states of the canonical commutation relations (I). Publ. RIMS. Kyoto Univ. 7 (1971/72), 105-120.
15. H. Araki, On quasifree states of the canonical commutation relations (II). Publ. RIMS. Kyoto Univ. 7 (1971/72), 121-152.
16. H. Araki and S. Yamagami, On quasi-equivalence of quasifree states of the canonical commutation relations. Publ. RIMS. Kyoto Univ. 18 (1982), 238-338.
17. C. C. Bernido and M.V. Carpio-Bernido, White noise functional approach to polymer entanglements. Proc. Stochastic Analysis: Classical and Quantum. Nov. 2004, World Sci. Pub. Co. 2005.
18. Ph. Blanchard, C.H.Chang and Krüger, Epidemic thresholds on scale-free graphs: the interplay between exponent and preferential choice. Ann. Henri Poincaré 4, Suppl. 2 (2003) 957-970.
19. S. Bochner, Harmonic analysis and the theory of probability. Univ. of California Press. 1955.
20. W.G. Cochran, H.-H. Kuo and A. Sengupta, A new class of white noise generalized functions. Infinite Dimensional Analysis, Quantum Probability and Related Topics. 1 (1998), 43-67.
21. P.A.M. Dirac, The principles of quantum mechanics. Oxford University Press. 1930.
22. P.A.M. Dirac, Lagrangian in quantum mechanics. Phys. Zeitschrift Sowjetunion. 3 (1933), 64-72.
23. J. L. Doob, Stochastic processes. Wiley & Sons. 1952.
24. A. Einstein, Einstein's 4 papers appeared in Annalen der Physik. 17 (1905), 132-148, 549-560, 891-921; 18 (1905), 639-641.
25. M. N. Feller, The Lévy Laplacian. Cambridge Tract in Math 166, Cambtridge Univ. Press, 2005.
26. W. Feller, An introduction to probability theory and its applications. Vol I. 1950; Vol II. 1966, Wiely.
27. M. Fukushima, Dirichlet forms and Markov processes. Kodansha and North-Holland. Tokyo and Amsterdam. 1980.
28. I. M. Gel'fand,
29. I.M. Gel'fand, Generalized random processes. (in Russian), Doklady Acad. Nauk SSSR, 100 (1955), 853-856.
30. I.M. Gel'fand and N.Ya. Vilenkin, Generalized functions, vol.4. Academic Press, 1964. (Russian Original, Moscow, 1961).
31. J. Glimm and A. Jaffe, Quantum physics. A functional integral point of view. Springer-Verlag. 1981.
32. M.B. Green, J.H. Schwarz and E. Witten, Superstring theory, I and II, Cambridge Univ. Press. 1987.
33. L. Gross, Potential theory on Hilbert space, J. Funct. Anal. 1 (1967) 123-181.
34. A. Hahn, Chern-Simonas theory on R^3 in axial gauge. Bonner Mathematische Schriften, Nr.345. 2001.
35. T. Hida, Canonical representations of Gaussian processes and their appli-

cations. Memoires Coll. of Sci. Univ. of Kyoto, A33 (1960), 109-155.
36. T. Hida, Stationary stochastic processes. Mathematical Notes. Princeton University Press. 1970.
37. T. Hida, Note on the infinite dimensional Laplacian operator. Nagoya Math. J. 38 (1970), K. Ono volume, 13-19.
38. T. Hida, Complex white noise and infinite dimensional unitary group. Lecture Notes in Math. Mathematics, Nagoya Univ. 1971.
39. T. Hida, A role of Fourier Transform in the theory of infinite dimensional unitary group. J. Math. Kyoto Univ. 13 (1973), 203-212.
40. T. Hida, Brownian motion. Springer-Verlag. 1980. Japanese Original: Buraun Unidou. Iwanami Pub. Co. 1975.
41. T. Hida, Analysis of Brownian functionals. Carleton Math. Notes no.13, 1975.
42. T. Hida, A note on stochastic variational equations. Exploring Stochastic Laws. Korolyuk volume. eds. A.V. Skorohod and Yu. V. Vorovskikh. 1995, 147-152.
43. T. Hida, Causal analysis in terms of white noise. Quantum Fields-Algebra, Processes. ed. L. streit, Springer-Verlag, 1980. 1-19.
44. T. Hida (with S. Albeverio, J. Potthoff, M.Röckner and L. Streit), Dirichlet forms in terms of white noise analysis. I: Construction and QFT examples. Rev. Math. Physics 1 (1990) 313-323.
45. T. Hida, Stochastic variational calculus. Stochastic Partial Differential Equations and Applications eds. B.L. Rozovskii and R.B. Sowers. Springer-Verlag. Lecture Notes in Math. 1992. 123-134.
46. T. Hida, A role of the Lévy Laplacian in the causal calculus of generalized white noise functionals. Stochastic Processes. G. Kallianur volume. ed. S. Cambanis. 1993, 131-139.
47. T. Hida, Random fields and quantum dynamics, Foundation of Physics. 29 (1997) Namiki volume. 1511-1518.
48. T. Hida, Some methods of computation in white noise analysis. Unconventional Models of Computation UMC'2K. eds. I. Antoniou et al. 2001, 85-93.
49. T. Hida, Innovation approach to stochastic processes and quantum dynamics. Foundations of Probability and Physics. 8 (2001), 161-169.
50. T. Hida, Complexity and irreveresibility in stochastic analysis. Chaos, Soliton and Fractals 12 (2001), 2859-2863.
51. T. Hida, White noise and functional analysis.(in Japanese) Seminar on Probability vol. 60. 2002.
52. T. Hida, White noise analysis: A new frontier. Volterra Center Pub. N.499, January 2002.
53. T. Hida, White noise analysis: Part I. Theory in Progress. Taiwanese Journal of Math. 7 (2003) 541-556.
54. T. Hida, Laplacians in white noise analysis. Contemporary Mathematics. 317 (2003), 137-142.
55. T. Hida, Some of future directions of white noise analysis. Proc. Abstract and Applied Analysis. ed. N.M. Chuong. (2003).
56. T. Hida, A frontier of white noise analysis in line with Itô calculus. Advanced

Studies in Pure Math. 41 (2004), 111-119.
57. T. Hida, Stochastic variational equations in white noise analysis. Proc. SPDE eds. G. Da Prato and T. Chapman, Chapman & Hall/CRC, (2005) 169-178.
58. T. Hida, Information, innovation and elemental random field Quantum Information and Computing, QP-PQ, eds. L. Accardi et al. vol.19 (2006), 195-203.
59. T. Hida and M. Hitsuda, Gaussian processes. Transl. Math. Monographs 120, American Math. Soc. 1993.
60. T. Hida and N. Ikeda, Note on linear processes. J. Mat. Kyoto Univ. 1 (1961), 75-86.
61. T. Hida and N. Ikeda, Analysis on Hilbert space with reproducing kernel arising from multiple Wiener integral. Proc. 5th Berkeley Sump. on Math. Stat. Probab.2 (1967), 211-216.
62. T. Hida, I. Kubo, H. Nomoto and H. Yoshizawa, On projective invariance of Brownian motion. Publ.RIMS, Kyoto Univ. A.4, (1969), 595-609.
63. T. Hida and H. Nomoto, Projective limit of spheres. Proc. Japan Academy.
64. T. Hida, H.-H. Kuo, J. Potthoff and L. Streit, White noise. An infinite dimensional calculus. Kluwer Academic Pub. Co. 1993.
65. T. Hida, N. Obata and K. Saito, Infinite dimensional calculus and Laplacians in terms of white noise calculus. Nagoya Math. J. 128 (1992), 65-93.
66. T. Hida, J. Potthoff and L. Streit, Dirichlet forms and white noise analysis. comm. Math. Phys. 116 (1988), 235-245.
67. T. Hida, J. Potthoff and L. Streit, Energy forms and white noise analysis. New Methods and Results in Nonlinear Field Equations. Lec. Notes in Phys. #347(1989), 115-125.
68. T. Hida and Si Si, Stochastic variational equations and innovations for random fields. Infinite Dimensional Harmonic Analysis, Transactions of a German-Japanese Symposium. eds. H. Heyer and T. Hirai. 1995, 86-93.
69. T. Hida and Si Si, Innovation for random fields. Infinite Dimensional Analysis, Quantum Probability and Related Topics. 1 (1998), 499-509.
70. T. Hida and Si Si, Innovation Approach to some problems in quantum dynamics. Garden of Quanta, Essays in Honour of Hirashi Ezawa, World Scientific Pub. Co., 2003, 95-103.
71. T. Hida and Si Si, Innovation approach to random fields: An application of white noise theory. World Scientific Pub. Co. Ltd. 2004.
72. T. Hida and L. Streit, Generalized Brownian functionals. Proc. VI International Conf. on Math. Physics. Berlin 1981, LN in Phys. 153 (1982), 285-287.
73. M. Hitsuda, Multiplicity of some classes of Gaussian processes. Nagoya Math. J. 52 (1973), 39-46.
74. E. Hopf, Statistical hydrodynamics and functional calculus. J. Rational Mechanics and Analysis. 1 (1952), 87-123.
75. K.Itô, On stochastic differential equations. Mem. Amer. Math. Soc. 4 (1951), 1-57.
76. K. Itô, Multiple Wiener integrals. J. Math. Soc. Japan 3 (1951), 157-169.
77. K. Itô, Stationary random distributions. Mem. Coll. Sci. Univ. Kyoto A, 28

Math. (1953), 209-223.

78. Y. Ito and I. Kubo, Calculus on Gaussian and Poisson noises. Nagoya. Math. J. 111 (1988), 41-84.
79. S. Janson, Gauissian Hilbert spaces. Cambridge Tracts in Math. 129, Cambridge Univ. Press. 1997.
80. G.W. Johnson and G. Kallianpur, Homogeneous chaos, p-forms, scaling and the Feynman integral. Trans. Amer. Math. Soc. 340 (1993), 503-548.
81. J. Jost, Riemanian geometry and geometric analysis. 3rd ed. Universitat, Springer-Verlag. 2002.
82. T. Kailath, The innovation approach to detection and estimation theory. IEEE Proc. May (1970) 680-695.
83. G. Kallianpur, Some ramification of Wiener's idess on nonlinear prediction. N. Wiener: Collected Works vol. III, ed P. Masani, The MIT Press, 1981, 402-424.
84. K. Karhunen, Über die Struktur stationärer zufälliger Funktionen. Arkiv för Mathematik, 1, no.13 (1950), 141-160.
85. A. N. Kolmogorov, Wienersche Spiralen und einige andere interessante Kurven in Hilbertscher Raum. Doklady Acad. Sci. URSS 26 (1940) 115-118.
86. Yu. G. Kondratiev and L. Streit, Spaces of white noise distributions: Constructions, descriptions, applications. I. BiBos Preprint, no. 510, 1992.
87. Yu. G. Kondratiev, J. Potthoff, L. Streit and W. Westerkamp, Generalized functionals in Gaussian spaces: The characterization theorem revisited. J. Functional Analysis 141 (1996), 301-318.
88. I. Kubo, Quasi–flows. Nagoya Math. J. Vol. 35 (1969), 1-30; Quasi–flows II, Additive functionals and TQ-systems. Nagoya Math. J. Vol. 40 (1970), 39-66.
89. I. Kubo and S. Takenaka, Calculus on Gaussian white noise. I - IV. Proc. Japan Academy 56 (1980), 376-380, 411-416; 57 (1981), 433-437; 58 (1982), 186-189.
90. I. Kubo and Y. Yokoi, A remark on the space of test random variables in the white noise calculus (preprint) 1987.
91. T. Kuna, L Streit and W. Westerkamp, Feynman integrals for a class of exponentially growing potentials. J. Math. Phys. 39 (1998), 4476-4491.
92. H. Kunita and S. Watanabe, On square integrable Martingales. Nagoya Math. J. 30 (1967), 209-245.
93. H. -H. Kuo, White noise distribution theory. CRC Press Inc. 1996.
94. R. Léandre, Stochastic Adams theorem for a general compact manifold. Rev. Math. Phys. 13 (2001), 1095-1133.
95. R. Léandre, Theory of distribution in the sense of Connes-Hida and Feynman path integral on a manifold. Infinite Dimensional Analysis, Quantum Probability and Related Topics. 6 (2003) 505-517.
96. R. Léandre and H. Ouerdiane, Connes-Hida calculus and Bismut-Quillen superconnections. Prépub. Institut Élie Cartan. 2003/no. 17.
97. Y.J. Lee, Analysis of generalized Lévy white noise functiomnals. J.of Functional Analysis. 211 (2004), 1-70.
98. P. Leukert and J. Schäfer, A rogorous construction of Aberian Chern-Simons

path integrals using white noise analysis. Reviews in Math. Phys. 8 (1996), 445-456.
99. P. Lévy, Sur le équations intégro-différentielles définissant des fonctions de lignes. Thëses. 1911, 1-120.
100. P. Lévy, Lecons d'analyse fonctionelle. Gauthier-Villars, 1922.
101. P. Lévy, Théorie de l'addition des variables aléatoires. Gauthier-Villars. 1937. 2ème ed. 1954.
102. P. Lévy, Processus stochastiques et mouvement brownien. Gauthier-Villars, 1948. 2ème ed. 1965.
103. P. Lévy, Problèmes concrets d'analyse foctionnelle. Gauthier-Villars. 1951.
104. P. Lévy, Random functions: General theory with special reference to Laplacian random functions. Univ. of California Pub. in Statistics. 1 (1953), 331-388.
105. P. Lévy, A special problem of Brownian motion, and a general theory of Gaussian random functions. Proceedings of the 3rd Bekeley Symp. on Mathematical Statistics and Probability. vol.II, Univ. of Californian Press. (1956) 133-175.
106. P. Lévy, Fonction aléatoire à corrélation linéaires. Illinois J. Math. 1, no.2 (1957), 217- 258.
107. P. Lévy, Jaques Hadamard, sa vie et son oeuvre - Calcul fonctionnel et questions diverses. La vie et l'Oeuvre de Jaque Hadamard. Monographie no. 16 de l'Enseignement mathématique. Imprimerie Kundig, Genéve, 1967. 1-24.
108. P. Lévy, Fonctions de lignes et équations aux dérivées fonctionnelles, Comm. XIII Congréss international d'histoire des sciences. Moscow, 1971,
109. P. Lévy, Quelques aspects de la pensée d'un mathématician. Blanchard, 1970.
110. J.L. Lions and E. Magenes, Non-homogeneous boundary value problems and applications. vol.1. Springer-Verlag, 1972.
111. B. Mandelbrot and J. Van Ness, Fractional Brownian motions, fractional noises and aplications. SIAM Rev. 10 (1968), 422-437.
112. B. Mandelbrot, Les objets fractals. Forme, hasard et dimension. 2ème ed. Flammarion. 1984.
113. B. Mandelbrot, Introduction to fractal sums of pulses. Lévy flights and related topics in physiscs. eds. M. F. Schlesinger, (1995), 110-123.
114. P. Masani and N. Wiener, Nonlinear prediction. H. Cramér volume. John Wiley & Sons, 1959, 190-212.
115. H. Matsusima, K. Okamoto and T. Sakurai, On a certain class of irreducible unitary represntations of the infinite dsimensional rotation group. I. Hiroshima Math. J. 11 (1981), 181-193; II loc. cit. 12 (1982), 385-397.
116. H. P. McKean, Brownian motion with a several-dimensional time. Theory of Prob. and Appl. 8 (1963), 335-354.
117. J. Mikusiński, On the space of the Dirac Delta-function. Bull. de l'Acad. Polanaise des Sci. Ser. Sci. Math. Astr. Phys. 14. (1966), 511-513.
118. A.S. Monin and A.M. Yaglom, Statistical hydrodynamics. vol. 1, Nauka. 1965; viol.2, Nauka. 1967. (in Russian).

119. M. Nakaoka, Homology of the infinite symmeteric group. Ann. Math. 73 (1961), 229-257.
120. D. Nualart, Stochastic integration with respect to fractional Brownian motion and applications. Contemporary Mathematics. 336 (2003), 3-39.
121. N. Obata, White noise calculus and Fock space. LNM no.1577, Springer-Verlag. 1994.
122. N. Obata and K. Saitô, Cauchy processes and the Lévy Laplacian. Non-commutability, Infinite Dimensionality and Probability at the Crossroad, eds. N. Obata et al. 2002, 360-373.
123. M. Ohya, Some aspects of quantum information theory and their applications to irreducible processes. Reports on Math. Phys. 27 (1989), 19-47.
124. M. Ohya and D. Petz, Quantum entropy and its use. Springer-Verlag. 1993.
125. R.E.A.C. Paley and N. Wiener, Fourier transforms in the complex domain. American Math. Soc. Collocu. Publ. 1934.
126. J. Potthoff and L. Streit, A characterization of Hida distribution. J. Functional Analysis. 101 (1991), 212-229.
127. J. Potthoff and M. Timpel, On a dual pair of spaces of smooth and generalized random variables, Potential Analysis 4 (1995), 637-654.
128. S.I. Resnick, Heavy-tail phenomina. Probabilistic and Statistical modeling. Springer Series in Operations Research and Financial Engineering, 2007.
129. S. Rhode and O. Schramm, Basic properties of SLE. Ann. Math. 161 (2005), 883-924.
130. G. Roepstorff, Path integral approach to quantum physics. An introduction. Text and monograph in physics. Springer-Verlag, 2nd ed. 1996. Chapt.3.
131. V. A. Rohlin, On the fundamental ideas of measure theory. Math. Sbornik 25 (1949), 107-150, AMS Engl. Trans Ser 1. Vol 10. (1962), 1-54.
132. Yu. A. F. Rozanov, Innovation processes. John Wiley & Sons, 1997.
133. K. Saitô, Stochastic processes associated with a sum of the Lévy Laplacians. Stochastic Analysis and Applications. 2000.
134. K. Saitô and A. Tsoi, Stochastic processes generated by functions of the Lévy Laplacian. Quantum Information II. eds. T. Hida and K. Saitô. World Scientific Pub. Co. 2000, 183-194.
135. M.B. Green, J.H. Schwarz and E. Witten, Superstring theory. II, Cambridge Univ. Press.1987.
136. C.E. Shannon and W. Weaver, The mathematical theory of communication. Univ. of Illinois Press. 1949.
137. T. Shimizu and Si Si, Professor Takeyuki Hida's Mathematical Notes, vol. I. 2004. (informal publication).
138. M.F. Shlessinger et al eds. Lévy flight and related topics in Physics. 1994 Nice Workshop. Springer-Verlag 1995.
139. Si Si, A note on Lévy's Brownian motion I, II, Nagoya Math. Journal 108 (1987) 121-130; 114 (1989), 165-172.
140. Si Si, Variational calculus for Lévy's Brownian motion. Gaussian random fields, Series of Probability and Statistics, Satellite Conference for International Math. Congress, World Scientific, 1990, 364-373.
141. Si Si, Variational calculus for Lévy's Brownian motion. "Gaussian random

fields" Ser. Probability and Statistics 1. (1991), 364-373.
142. Si Si, Integrability condition for stochastic variational equation. Volterra center Notes N.217, 1995.
143. Si Si, Innovation of some random fields. J. Korea Math. Soc. 35 (1998), 575-581.
144. Si Si, A variational formula for some random fields; an analogy of Ito's formula. Infinite Dimensional Analysis, Quantum Probability and Related Topics. 2 (1999), 305-313.
145. Si Si, Random fields and multiple Markov properties, Supplementary Papers for the 2nd International conference on Unconventional Models of Computation, UMC'2K, ed. I. Antoniou, CDMTCS-147CCenter for Discrete Mathematics and Theoretical Computer Science, Solvay Institute, 2000, pp 64-70.
146. Si Si, Gaussian Processes and Gaussian random fields. Quantum Information U, Proc. of International Conference on Quantum Information, World Scientific. 2000, 195-204.
147. Si Si, X-ray data processing. - An example of random communication systems, Nov. 2001:
148. Si Si, Effective determination of Poisson noise. Infinite Dimensional Analysis, Quantum Probability and Related Topics. 6 (2003), 609-617.
149. Si Si, Poisson noise, infinite symmetric group and a stochastic integral based on quadratic Hida distributions. Preprint.
150. Si Si, A. Tsoi and Win Win Htay, Invariance of Poisson noise. Proceedings: Stochastic Analysis: Classical and Quantum. ed. T. Hida, World Sci. Pub. Co. 2005, 199-210.
151. Si Si, Note on fractional power distributions. (preprint)
152. Si Si and Win Win Htay, Entropy in subordination and Filtering. Acta Applicandae Mathematicae (2000) Vol. 63 No. 1-3 ,433-439
153. Si Si and Win Win Htay, Structure of linear processes. Quantum Information and and computing. eds. L. Accardi et al, 2006. 304-312.
154. Si Si, A. Tsoi and Win Win Htay, Jump finding of a stable process. Quantum InformationV, World Scientific Pub. Co. (2006) 193-202.
155. B. Simon, The $P(\phi)_2$ Euclidean (quantum) field theory. Princeton Univ. Press. 1974.
156. B. Simon, Functional integration and quantum physics. 2nd ed. AMS Chelsea Pub. 2005.
157. Ya.G. Sinai, Dynamivcal systems with countably multiple Lebesgue space II. Amer. math. Soc. Transl. (2) 68 (1968), 34-68.
158. F. Smithies, Integral equations. Cambridge Univ. Press. 1958.
159. E.M. Stein, Topics in harmonic analysis. Related to the Littlewood-Palay theory. Ann. Math. Studies. 63, Princeton Univ. Press. 1970.
160. L. Streit and T. Hida, Generalized Brownian functional and Feynman integral. Stochastic Processes and their Applications. 16 (1983), 55-69.
161. S. Tomonaga, On a relativistically invariant formulation of the quantum theory of wave fields. Progress of Theoretical Physics. vol.1 (1946), 27-39.
162. P. Topping, Repulsion and quantization in almost harmonic maps, and asymptotics of the harmonic map flow. Ann. Math. 159 (2004), 405-534.

163. N. Wiener, Generalized harmonic analysis. Acta Math. 55 (1930), 117-258.
164. N. Wiener, The homogeneous chaos. Amer. J. Math. 60 (1938), 897-936.
165. N. Wiener, Extrapolation, interpolation and smoothing of stationary time series. The MIT Press, 1949.
166. N. Wiener, Nonlinear problems in random theory, The M.I.T. Press. 1958.
167. Win Win Htay, Markov property and information loss. preprint, Nagoya Univ.
168. E. Witten, Quantum field theory and the Jones polynomial. Commun. Math. Phys. 121 (1989), 353-389.
169. Y. Yokoi, Positive generalized functionals. Hiroshima Math. J. 20 (1990), 137-157.
170. Y. Yokoi, Simple setting for white noise calculus using Bergmann space and Gauss transform. Hiroshima Math. J. 25 (1995), 97-121.
171. H. Yoshizawa, Rotation group of Hilbert space and its application to Brownian motion. Proceedings of International Conference on Functional Analysis and Related Topics. Tokyo (1970), 414-423.
172. K. Yosida, Functional analysis. Springer-Verlag, 6th ed. 1980.

Subject Index